FLEAS

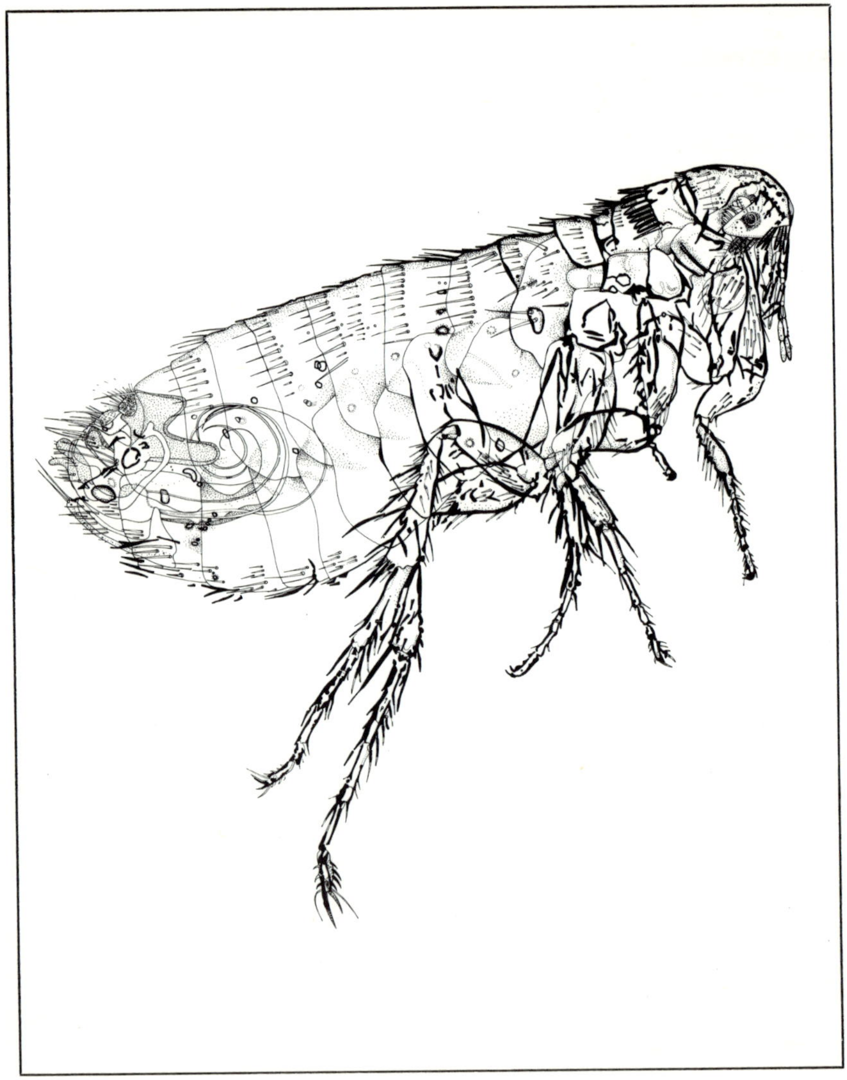

FLEAS

BY HELEN HOKE
AND VALERIE PITT

FRANKLIN WATTS, INC.
NEW YORK, 1974

A FIRST BOOK

First published in England by Franklin Watts, Ltd.

Acknowledgments
We would like to thank Mr. F.G.A.M. Smit, custodian of the Rothschild and British Museum (Natural History) collections of fleas, for reading our manuscript and for his helpful suggestions.

Perhaps, most of all, we owe thanks to a gentleman whom we have never met–Mr. Brendan Lehane. For it was his witty book *The Compleat Flea* which first turned *us* on to fleas.

Helen Hoke and Valerie Pitt

Credits
"The Flea" by Sir Alan Herbert, reprinted by permission of Lady Herbert; Line drawings by James Robins; Photos by John Drysdale, copyright Camera Press Limited, London: page 51, pages 54-55; The British Museum: page 58; Copyright Radio Times Hulton Picture Library: page 61, page 64, page 72 bottom; New York Public Library Picture Collection: page 68, page 72 top.

Library of Congress Cataloging in Publication Data

Hoke, Helen, 1903-

Fleas.
(A First book)
Bibliography: p.
SUMMARY: Describes the anatomy and habits of fleas and how the diseases they carry have, in some cases, altered history.
1. Fleas–Juvenile literature. [1. Fleas] I. Pitt, Valerie, joint author. II. Title.
QL599.5.H64 595.7'75 73-12147
ISBN 0-531-00815-0

Printed in the United States
5 4 3 2 1

CONTENTS

"I never know why it should be
So rude to talk about the flea.
What funny folk we are.
I think we've got the jealous hump
Because we know we'll never jump
So skilfully and far."

Sir Alan Herbert

1

THE PLAGUE-Y FLEA

Too rude to talk about. An embarrassment. A joke. A menace. A circus "artiste." A killer. In its time the tiny flea has been all these things.

It has also been an unwitting troublemaker with a wry sense of humor.

A horde of fleas once caught a singer in the spotlight at a Paris opera. She started to scratch — then could not stop. Soon the audience was scratching along with her.

A compartment on the famous Orient Express emptied rapidly on one journey. The conductors wondered what had happened until they found that an Indian passenger had dropped a box containing three hundred performing fleas.

A late-night sitting of Belgium's Parliament was once disrupted by fleas. All the benches kept emptying as embarrassed members went out to scratch. The president himself had to leave the rostrum three times.

Fleas have even been the cause of a lawsuit. A man who went on vacation to Spain found fleas in the bed in his hotel room. He sued the travel company because his family's holiday had been ruined. The judge on the case commented, "They seem to have paid a lot to sleep in the society of fleas."

Fleas once caused German customs officials a bit of trouble. An English flea circus owner, on tour with her troupe, had declared her performers as "insects." Nonsense, said the officials with their customary passion for correctness, the fleas were "artistes" and must be taxed on each performance. "We even had to make out a report if one died and also if we took on a new one," said the owner.

The Communist youth paper in Prague, Czechoslovakia, decided it was against the whole idea of flea circuses. It told a Czech flea circus owner to free his fleas and get down to some honest work. "Flea circuses are not needed in a Socialist democracy," said the paper.

A few years ago some fleas almost made it into space. They were found in a supposedly germproof test chamber which was part of an orbiting laboratory. The temperature in the chamber was kept at such an icy level, minus 76 degrees F., that scientists who looked inside had to wear special facemasks and fur-lined parkas. Perhaps, conjectured the bewildered scientists, the fleas had dropped from the parka of a previous visitor to the chamber?

Fleas were again the culprits when an office in Gloucestershire, England found it had a strike on its hands. A cleaner had been "bitten" by a flea and nineteen other cleaners walked out in sympathy with her. Next day the crisis was over. Paper waste sacks were brought in to replace some burlap ones which had proved happy breeding grounds for the fleas.

Yet, despite all the trouble they have caused, fleas have also been harbingers of pleasure. In 1951 a Major Robert Traub described a flea which a colleague had collected from a rat found scuttling about on a mountain slope in the Philippines. Later, at a meeting in New York City, Major Traub presented a beautifully sculptured larger-than-life-size model of this flea to Mr. Charles Pomerantz, a former New York dress designer and an expert on parasites.

Said Major Traub, "I am naming this species *Stivalius pomerantzi* after Mr. Pomerantz, who by his studies has contributed so much to science."

Said Mr. Pomerantz, "I feel a real and deep humility in having this flea called by my own name."

2

THE LIST OF CHARGES

What is it about fleas that has caused so much controversy? Why are fleas *rude*? Why are they *funny*? Why a *menace?* Why *killers*? What, indeed, *is* a flea?

From the very beginning fleas have had two major strikes against them: they are insects — and they are parasites. Neither characteristic endears them much to man.

And, then, if you've got them, they *itch.* And scratching — like speaking with your mouth full — is *rude.* It brings to mind too many things that people don't like to talk about, too many things associated with the popular image of horrid little sucking creatures. Like dirt. And houses full of dust and rubbish. And unwashed clothes. And smells. No one likes to think he's *dirty*. It's *embarrassing.* And when people get embarrassed they often turn the cause of their discomfort into a *joke*. There are lots of jokes about fleas.

And then again, far more serious charges can be laid against the flea. Some are carriers of disease — one of which is so gruesome that even its name carries a sort of dread: the plague. Plague has been responsible for more deaths than all the bombs and guns and ammunition of World War II.

So, fleas have a lot to answer for.

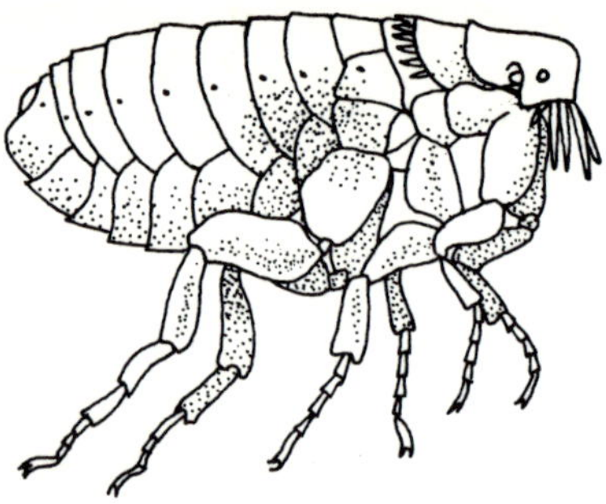

Flea

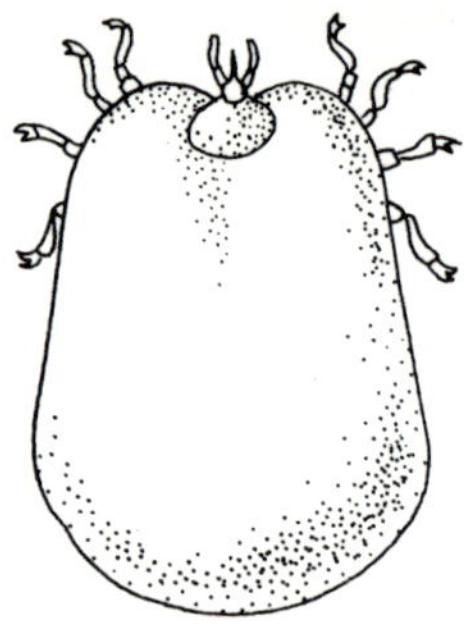

Tick

Bedbug

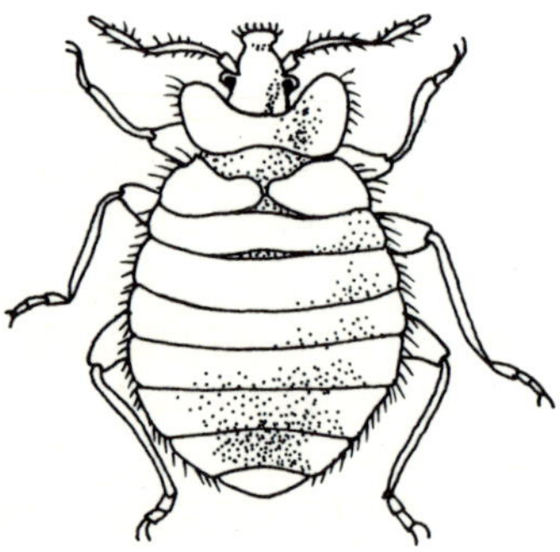

Louse

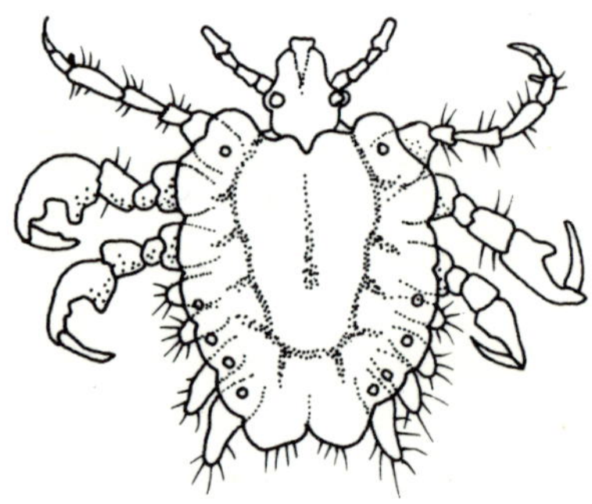

3

WHAT THEY'RE NOT

Although some of the world's shudders can be traced to fleas, they are not responsible for all the horrors that people might attribute to them. Fleas are often confused with a whole host of other blood-sucking creatures and sometimes take the blame for their misdeeds. So, first, let's decide what fleas are not.

They are not lice. There are three kinds of lice which cling to man and suck his blood. The most well-known are head lice. The eggs of head lice are called "nits" and people often confuse them with fleas. Another kind of lice are body lice. People may catch body lice when they are living in overcrowded conditions with little opportunity to wash or change their clothes. During wartime, for instance, these lice can be great pests. Besides causing intolerable itching, they can also carry a serious disease — typhus.

People have been afflicted by lice for centuries. When Thomas à Becket was murdered in the twelfth century scores of lice crawled out of his clothes. During the seventeenth and eighteenth centuries, fashionably dressed men and women often carried long "head scratchers" like elaborate knitting needles, to attack the lice infesting their wigs. There is even a story that during the Middle Ages a louse was used to "elect" the new mayor of a village in Sweden. All the eligible men were seated around a table, then they bowed their heads so their beards touched the top of it. A louse was placed in the middle of the table and the person whose beard it entered was elected to office.

Fleas are not mites or ticks, either. The most well-known of these tiresome creatures is the harvest mite which during the early stages of its development inserts its tiny fangs into any warm-blooded animal that happens to be around. It is often found in chalky areas and gardens. When people start to itch after a picnic in the open air, these mites are often responsible.

Nor are bedbugs fleas. These bloodsucking insects produce a foul smell and infest dirty houses. They lay their eggs in cracks and crevices and come out at night to search for "food."

4

"BUGS" AND INSECTS

So, what exactly is a flea? First and foremost, a flea is an insect. "Insect" means many things to many people: some people call all small, creepy-crawly things "bugs" and leave it at that. Others call spiders and scorpions *insects*, which is enough to curl the hair of any self-respecting scientist. What, then, *are* insects?

Along with five other classes of living creatures whose bodies are hardened on the outside and divided into segments, insects belong to the major group of animals known as *Arthropoda.* Within this group they have their own classification —*insecta.* Spiders, scorpions, mites, and ticks have their own classification too — *arachnida.* And crabs, lobsters, and shrimp have theirs — *crustacea.*

There are two main ways of telling the difference between insects and other arthropods like crustacea and arachnida. The first is this: if you remember that "insect" comes from the Latin word *insectum*, meaning *cut into*, you can easily recognize an insect by the way its body is "cut" into three main divisions — the head, the thorax, and the abdomen. And the second is this: all insects have three

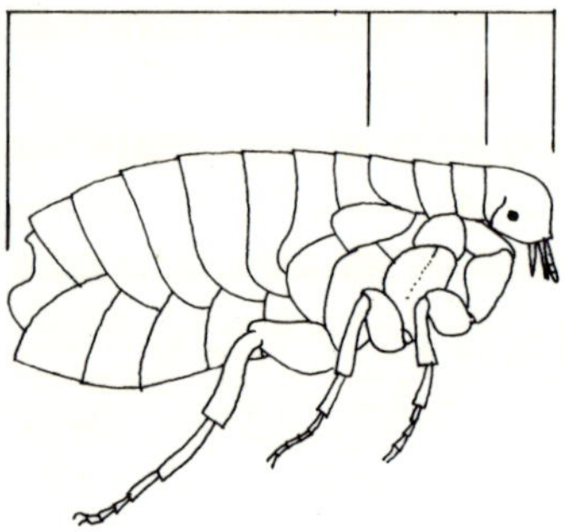
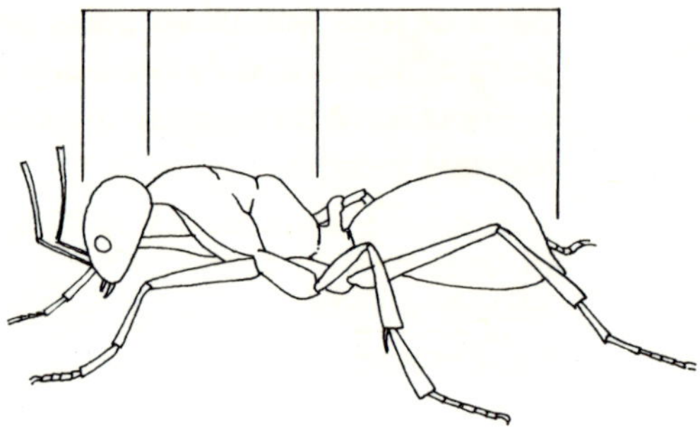
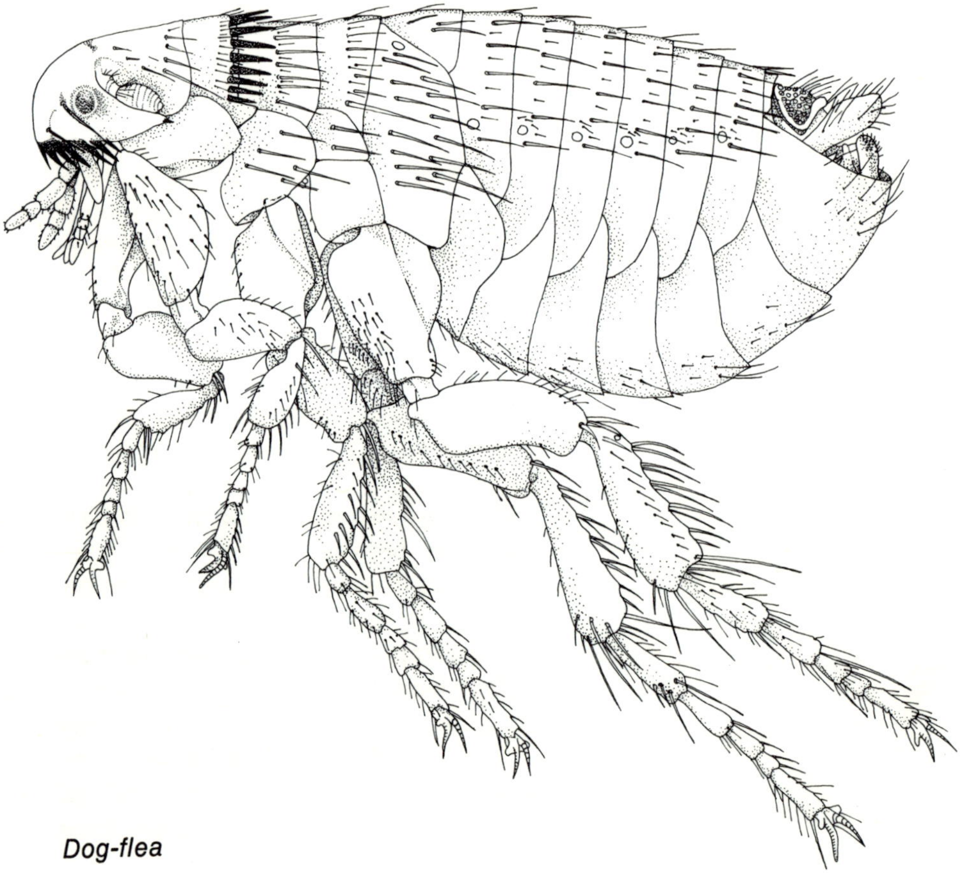

Dog-flea

pairs of legs and three pairs only. (Spiders, for instance, have four pairs of legs and only two body divisions.)

Fleas have three body divisions and three pairs of legs; therefore they are insects.

The 2,000-plus known species and subspecies (or different kinds) of fleas belong to the order *siphonaptera* (sigh-fon-a-pter-a), which means *wingless sucking insects.* But this relatively small order of fleas makes up only a tiny fraction of the enormous class of insects. There are at least 800,000 different species of insects in the world, from all the delicate varieties of butterflies to all the different kinds of beetles. That's almost three times the number of *all* the other known species of animals which inhabit the earth. Some experts believe that when all the yet-to-be-discovered species of insects are counted up the total number will reach a million or more.

This vast assembly of flying, jumping, crawling, buzzing, humming creatures circles the globe. Insects live in deserts, jungles, mountains, steppes, and plains; some even spend their lives in the forbidding icy world of Antarctica. Just about the only place where few insects live is the sea and then only around the shore, but ponds and streams may be alive with them.

Many of these millions of insects act as nature's "refuse workers." Some feed on each other, helping to keep down the teeming population. Some feed on dead matter, clearing the way for new life. Without insects the earth would eventually suffocate, piled high with the unburied bodies of animals and dead plants.

5

THE "GOOD" AND THE "BAD"

To the scientist there are no "good" or "bad" insects. A scientist will tell you that insects are just "here" as *we* are "here" and the Atlantic Ocean is "here," and the fact that they do the vital job of clearing up the earth is just a happy coincidence for man.

But man, with his love of labeling, likes to classify animals, particularly as they relate to him. To man, there *are* "good" insects and "bad."

Under the heading of "good," of course, come those which are commercially profitable—the silkworm moth, the honeybee, the lac insect, from which we get silk, honey, and furniture polish. Bees and other insects which carry pollen from one plant to another, ensuring a mass of blossoming flowers, are "good" too.

The list of "bad" insects is much longer. Pretty well all the other insects are lumped under this heading. These are the "enemies" of man: the fleas, cockroaches, bedbugs, and beetles who live with him in the cracks and crannies of his houses. The moths whose larvae chew up his clothes. The termites who chew at his buildings. The grasshoppers, caterpillars, locusts, and sawflies who devour his crops. The fleas, mosquitoes, lice, and others who pass on to him some of the world's most terrible diseases.

The flea comes very high on the list of the unloved. For besides being an insect, it is also that most despised of creatures — a parasite.

6

HOSTS AND PARASITES

There is a tiny worm which dwells under the eyelid of a hippopotamus, feeding on a very exclusive diet — the hippo's tears.

There is another which lives in the veins of a seagull, gorging on its blood.

A tapeworm, as long as a cricket pitch (22 yards) may spend some of its life in the intestines of man.

A flea may spend a good deal of its adult life on another animal, piercing its skin and then sucking up its blood for food.

All of these creatures are *parasites.* They live at the expense of others, feeding either inside another live animal or on it, without giving any benefits in return. (There are parasites of plants, too.) It is probably only natural that human beings should despise parasites — especially when they settle on their own bodies — in the same way that they despise people who live off society without contributing to it, for such lives are a one-way street.

Fleas are *ecto-parasites.* They feed *on* the bodies of other animals, not inside them like the *endo-parasitic* worms.

Any creature, willing or not, which harbors a parasite and provides its meals, is called a *host.* Many smallish mammals,

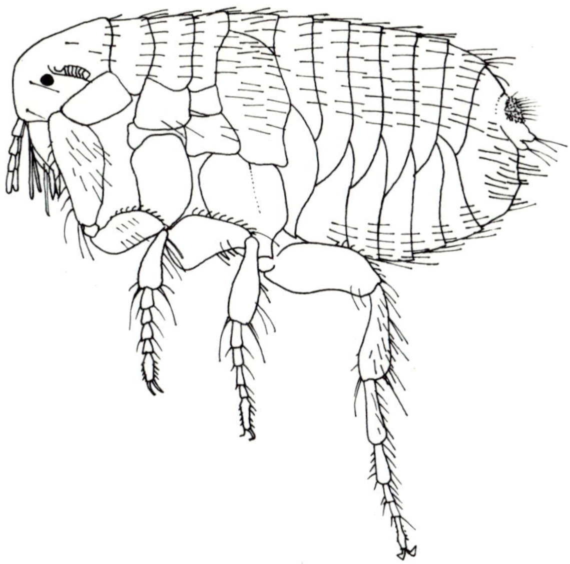

Rat-flea

like dogs, cats, rats, squirrels, rabbits, badgers, moles, and even bats, are hosts to certain species of fleas. Sand martins, hens, chaffinches, sparrows, and other kinds of birds, are hosts to some other species of fleas. And human beings — perhaps you, perhaps us, perhaps "them" — are hosts to yet another. Most of these animals are hosts to specific kinds of fleas, which are named after them. Thus, there are cat-fleas, dog-fleas, human-fleas, and many others.

Sometimes, when fleas cannot find their natural hosts and begin to grow hungry, they will switch their affections for awhile. A naturalist once collected more than a thousand cat- and dog-fleas from human habitations like concert halls, ballrooms, and barracks. Hen-fleas and rat-fleas will also alight on humans if their natural hosts are not around. In cases like this we human beings are the *accidental hosts.*

Monkeys are not natural hosts to fleas. All that scratching you see at the zoo is to clear their coats of dirt and dandruff. But monkeys can become accidental hosts by catching human-fleas from people crowding around their cages!

So. The flea is an insect. And a bloodsucking parasite. It is also a most extraordinary looking creature.

7

"PINCHED" INTO SHAPE

To the naked eye a flea looks like a tiny brownish wriggling grain of rice, barely distinguishable from its surroundings. Most are not much bigger than one-tenth of an inch long; even Great Britain's largest known flea, the mole-flea, is only about one quarter of an inch long. And fleas weigh next to nothing, perhaps .00071 grams or as much as the dot of the letter i.

A human-flea about life size

It is not until you see a flea under a microscope, or look at a greatly enlarged picture of a flea, that you can detect its amazingly complex structure and shape. A flea is truly "the world in a grain of sand."

The first thing you notice about a flea is that it is strangely flattened on both sides, as though something much larger and stronger had come along and "pinched" it into shape. This makes it taller than it is broad. Then, it is streamlined from head to tail, without neck or waist to define its body like most other insects. And, covering its body, are backward-pointing spines or bristles arranged in rows and groups.

A flea travels in different territory from most other insects; through the fur, feathers, or hair of its host. Its body has evolved to suit its unusual lifestyle, to help it navigate as easily as possible.

Its shape, streamlined body, and spines all help the flea to glide through often tangled terrain.

8

WATERPROOF "ARMOR"

All insects have external skeletons, called *exo-skeletons*, for protection— not only from enemies but also against the loss of heat and moisture. Hard and shiny, the flea's skeleton is made up of a horny waterproof substance called *chitin* which covers its entire body like a suit of armor. But it is flexible armor. The chitin varies in thickness and is much thinner at the joints to allow for movement.

The flea's head, which is usually pointed or rounded to help it make a path through the host's fur, is generally smooth. The rest of its body is covered in a series of loosely overlapping plates, like the "shields" of a shrimp. Some species of fleas have tiny chitinized "combs" — like ruffs — on their heads and bodies which may help protect them from getting jabbed in their joints by the sometimes spiky hairs of their hosts. Hedgehog-fleas do not have developed combs. One would imagine they would suffer torment from the bristles of their hosts, but actually hedgehog hairs are coarse and probably not very irritating to fleas and they certainly do not stop hedgehog-fleas from congregating in large num-

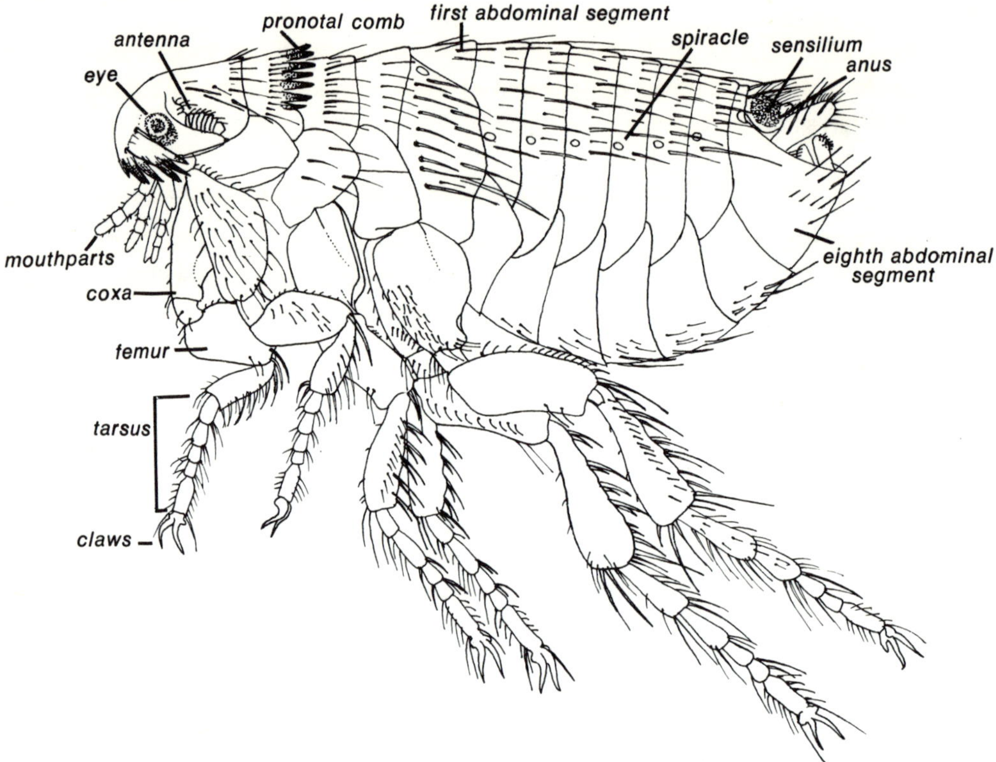

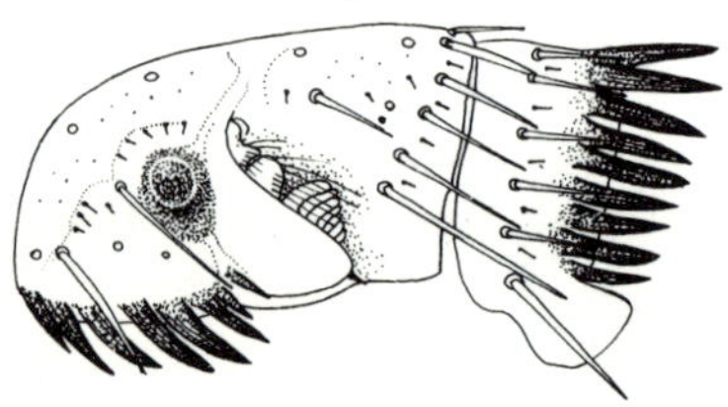
Dog-flea

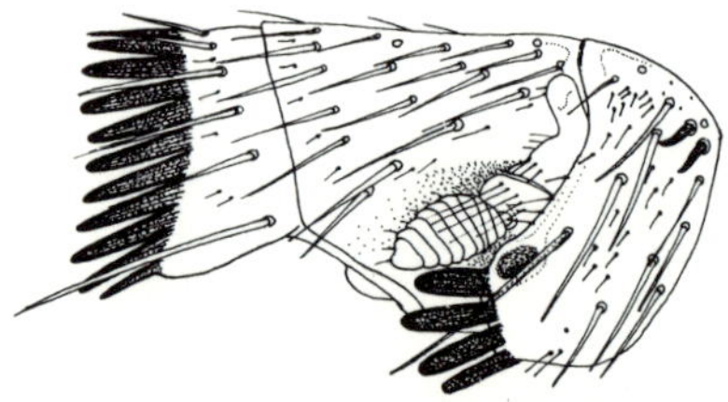
Mouse-flea

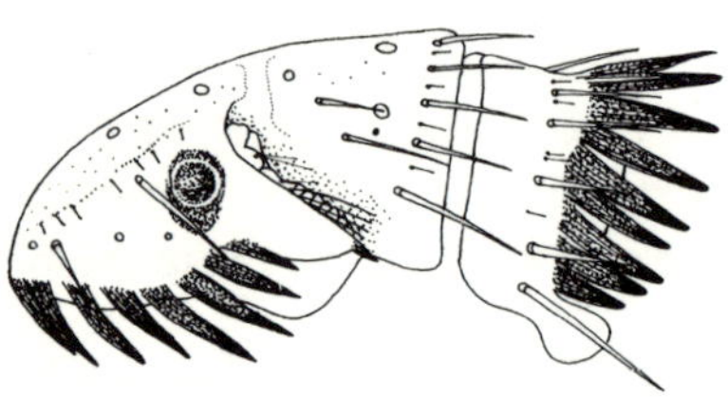
Cat-flea

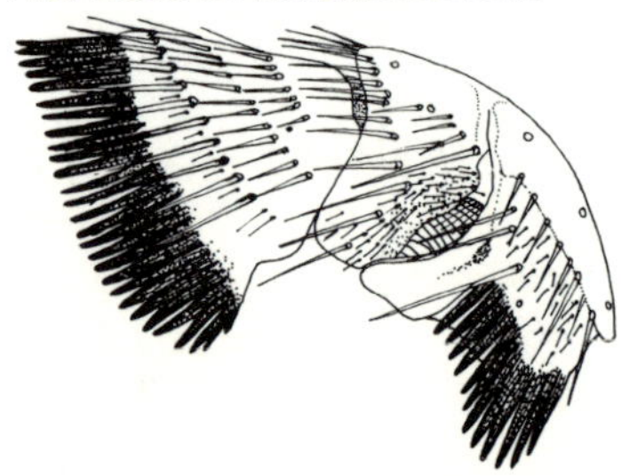
Flea found on bats and rodents

bers. Mr. F.G.A.M. Smit, Custodian of the Rothschild and British Museum (Natural History) collections of fleas, once counted 932 fleas on one young hedgehog.

Whole flea and heads of various types showing different development of combs

9

HEAD FIRST

The head. The thorax. And the abdomen. These are the three main body units of fleas and all other insects. They are connected to each other by flexible joints.

The flea's tiny head contains its mouthparts, eyes (if any), and antennae. Before a flea can suck up blood, it must first find a spot on its host's skin and pierce it. Two hard chitinous plates on either side of its face help the flea to separate the hairs of its host and find the bare skin. Then its *proboscis* — a sort of tube made up of three stiletto-shaped blades — pierces the host's skin and siphons up the blood. The proboscis is usually a very delicate instrument, so the piercing does not alert the host.

Fleas which are parasites of burrowing creatures inhabiting dark places or only coming out at night seldom have developed eyes — they would be of little use. Those which are parasites of diurnal, or daytime, creatures generally do have eyes but their vision is limited, perhaps just useful for detecting the difference between light and shade. Fleas use their sense organs rather than eyes to find their way about. They have two pairs of touch organs, called *palps*, in front of their mouthparts and a pair of antennae placed just behind their eyes.

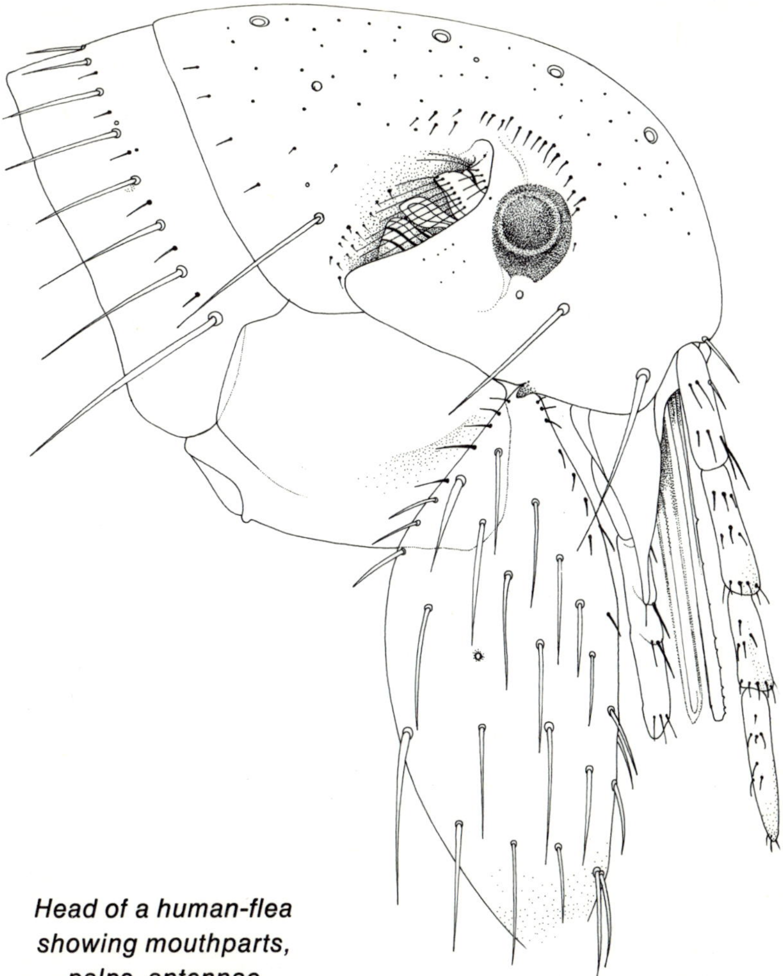

Head of a human-flea showing mouthparts, palps, antennae

The antennae, organs of touch and possibly smell, help keep the flea in contact with the world. Male fleas also use their antennae to clasp the females during mating. When not in use the antennae of fleas, unlike those of moths or grasshoppers, rest in grooves. This is another adaptation to keep the flea as streamlined as possible when it is crawling through the hair of its host.

10

JUMPING OVER THE WASHINGTON MONUMENT

After the head comes the thorax or "chest," which is divided into three segments. Attached to the underside of the thorax are the three pairs of legs — a pair to each segment.

Fleas do not have wings like most other insects, for wings would impede their progress through hair or fur. Instead they use their legs for locomotion. Fleas hop, run, walk — and, above all — *jump.* For their size and weight, fleas are among the strongest jumpers in the animal world, far more adept than locusts, frogs, or kangaroos. A human-flea can jump as high as seven to nine inches. That's more than eighty times its own length. If man had such dazzling powers he could almost make it over the top of the Washington Monument! Horizontally, a flea can cover close to a foot in one leap.

Most of the power for these jumps comes from the rear pair of legs, which are strongly jointed and larger than the others. Just above these hind legs, in the thorax, is a small depression which holds a colorless protein called *resilin*, capable of storing and releasing energy. As the flea starts to jump, pressure releases the energy stored in the resilin and propels the flea forward.

Because each of their legs ends in a pair of claws, fleas cannot walk or run easily on smooth surfaces. The claws are an adaptation to help them grip their hosts.

11

BREATHING THROUGH TUBES

The third section of the flea's body is its abdomen, which is divided into ten segments. Here lie all the most important organs of life — the heart, the digestive, excretory, and reproductive organs. It is here, too, that the flea does most of its breathing.

Fleas do not have lungs like human beings, neither does their blood absorb oxygen and carry it to the tissues like ours does. Fleas breathe through tiny openings in their abdomen and thorax, called *spiracles.* Air enters the spiracles and is carried to all parts of the body by an elaborate system of branching tubes.

Far back on the abdomen lies one of the flea's most important organs: a tiny sensory plate, covered with hair, called the *sensilium.* Air currents, heat and cold, and vibrations are all picked up by the sensilium. It is by following these stimuli that the flea finds a host, or during breeding times — a mate.

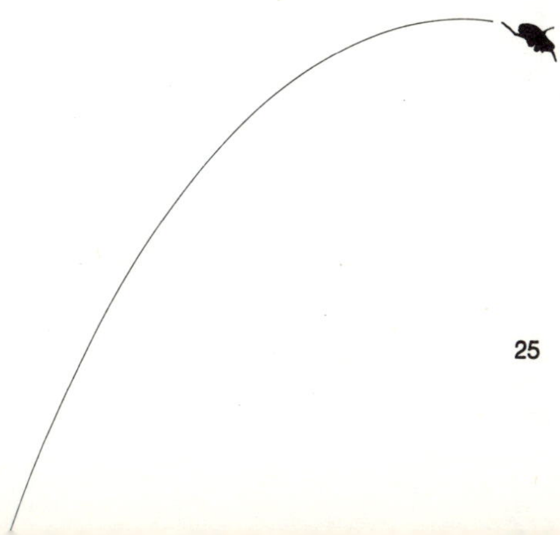

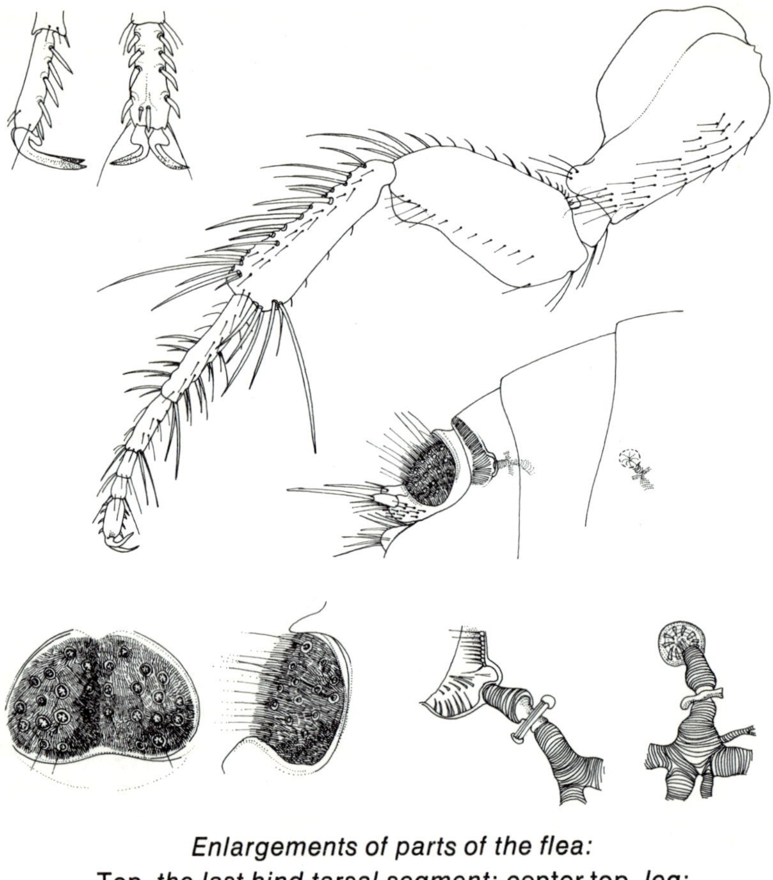

Enlargements of parts of the flea:
Top, *the last hind tarsal segment*; center top, *leg*; center below, *rear of flea showing sensilium and spiracles*; bottom left, *sensilium enlarged*; bottom right, *spiracle enlarged*

12

CHANCE MEETINGS

In order to live a flea must feed and in order to feed it must find a host. Its success or failure depends on whether it is in the right place at the right time.

The place where a flea stands most chance of finding a host is in the nest or burrow where the host makes its home, and this is in fact where many fleas live and breed: the young emerging from their cocoons with (if all the conditions are favorable) a plentiful food supply nearby.

To this end, those fleas which are parasites of mammals have been much more successful than those species which are parasites of birds. Of the known species of fleas about 94 percent are parasites of mammals and only about 6 percent are parasites of birds.

Birds build nests for the hatching and feeding of their young and when these jobs are accomplished they may fly off to new pastures leaving the fleas and other parasites to starve till the beginning of a new nesting season, unless they can manage to hitch a lift on another passing host. Mammals like rats, moles, cats, dogs, and man have much more permanent homes: a nest, a burrow, a basket, a kennel, a house. These are the sort of creatures which are most afflicted by fleas, for in order to tap the food supply fleas must live in those permanent homes too. There is one unusual mammal which does not make a nest but is nevertheless a host to fleas — the bat. This is explained by the fact that bats do return to the same place, or roost, to sleep. Cattle and horses are more nomadic animals and are only infested in certain parts of the world by specialized fleas.

Despite the fact that most fleas are parasites of hosts which make nests, finding a host is not always easy. It is all a tremendous game of chance and many fleas die within the first month or so of life. The flea has only a few tiny attributes to help it find a host — the palps, the antennae, and the sensilium — and the numbers are stacked against it, as they are against all other parasites. Just imagine all the millions of parasites in the world waiting for that one coincidence which will bring them close enough to make contact with a passing host. And fleas may only be able to grip a host with their claws if it passes within a few inches. Even very hungry fleas may wander around aimlessly at first if they are put into a cage with a rat. It may be a few hours before the majority finally latch onto it.

Experts believe that fleas rely to a great extent on their sense of smell when they are looking for a host. Rabbit-fleas, for instance, are often attracted by the urine of their hosts. But no one is exactly sure just which organ is responsible for smell — it may be the antennae, it may be the palps, it may be both. But "sniff" fleas do, with a sense so acute that they can distinguish between different animals. The smell of horses seems to repel them in some way, and grooms and other people who handle horses a good deal are often immune to fleas. Some species of fleas will starve rather than suck blood from an accidental host.

Fleas are very sensitive to air currents, vibrations, and temperature, and it is by following these stimuli as well as scent — often in a laborious zig-zag fashion — that they find and manage to jump onto a host. A number of fleas were once collected by a scientist who dragged a hot water bottle, wrapped in a bird skin, through some undergrowth. In this case the fleas had scent, warmth, and vibrations to guide them.

13

BLOODSUCKING

By chance, or by laboriously following trails of scent and vibration, the fortunate flea has found a host. Now it begins to feed. Most fleas have voracious appetites and if left undisturbed may feed for hours at a time, even though they digest only a fraction of the blood they suck. But if the opportunity for feeding is not there, fleas are capable of fasting for long periods. They can go weeks and even months without food, but usually only when it is fairly cold.

Once it has cleared a spot on the host's skin, the flea sets to work. Raising its hind legs and abdomen in the air, it pushes its proboscis down hard to pierce the skin. Immediately, saliva starts to flow down through a thin gutter in the proboscis and into the wound. The flea, like most other bloodsucking parasites, has a special sort of enzyme in its saliva which acts on the blood of the host and prevents it from clotting. The saliva causes irritation to the host but from the flea's point of view it is a lifesaver; it stops the proboscis and the flea's intestines from becoming clogged with clotted blood.

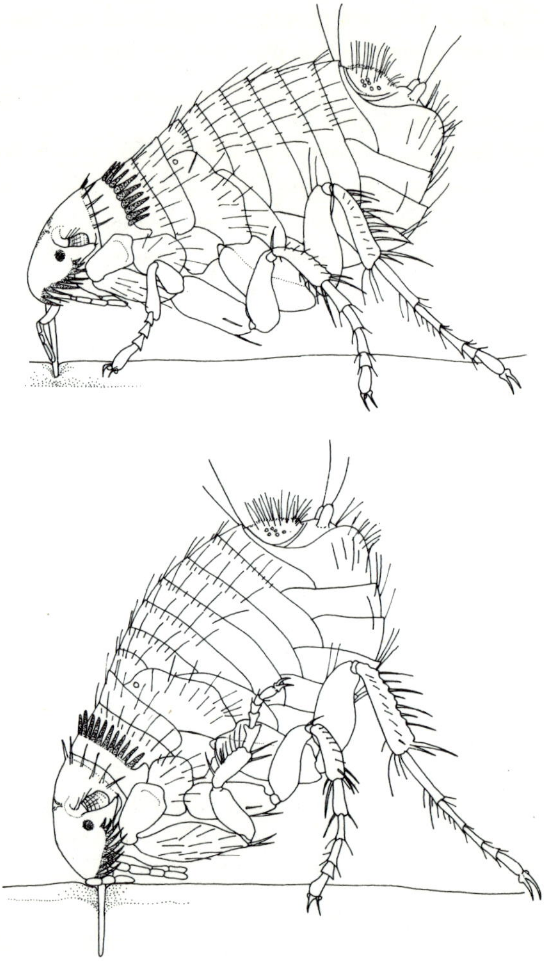

Feeding flea:
upper, *in probing position*; lower, *in sucking position*

Although the flea may spend hours swaying on its host's skin, it does not take in large amounts of blood. It is too small. And only a fraction of the tiny amount of blood siphoned up the proboscis is digested. The rest passes straight through the flea unchanged and spurts out from the anus in a trickle of drops. Often some of these drops land on the host's skin.

Unlike endo-parasites, which live inside their hosts, fleas are fairly independent parasites. Most only rely on their hosts for food. Once the host has provided the free meal, the flea will drop off and settle down close to or in the nest or burrow of its host, or, in the case of man, in the cracks and crevices of his houses. There it will bide its time until its next meal and take care of other tasks — digestion, excretion, and perhaps reproduction.

14

SOME SPECIALISTS

A few species of fleas live more dependent lives, attaching themselves almost permanently to their hosts. Rabbits often make temporary burrows and are up and off after a month or so. Rabbit-fleas may go along for the ride, following their food supply. Female rabbit-fleas are among the most dependent specialists. In fact, the life of the female rabbit-flea has become so intimately bound with that of its host, that it is dependent on the female rabbit to start its own reproductive cycle.

A female rabbit-flea can only breed if it has taken in the blood of a female rabbit. But more than that — the rabbit must be pregnant. The blood content of a rabbit changes when it is pregnant and the female rabbit-flea must take in some of this changed blood before it too can mate and lay eggs. On the last day of the rabbit's pregnancy the flea's eggs become ripe and ready for fertilization. A few hours after the young rabbits are born the female rabbit-flea switches over to one of them and imbibes more blood. After this meal it is ready for mating. If mating takes place then the rabbit-flea will soon start to lay her eggs.

This sort of specialization has advantages for the new fleas when they emerge,

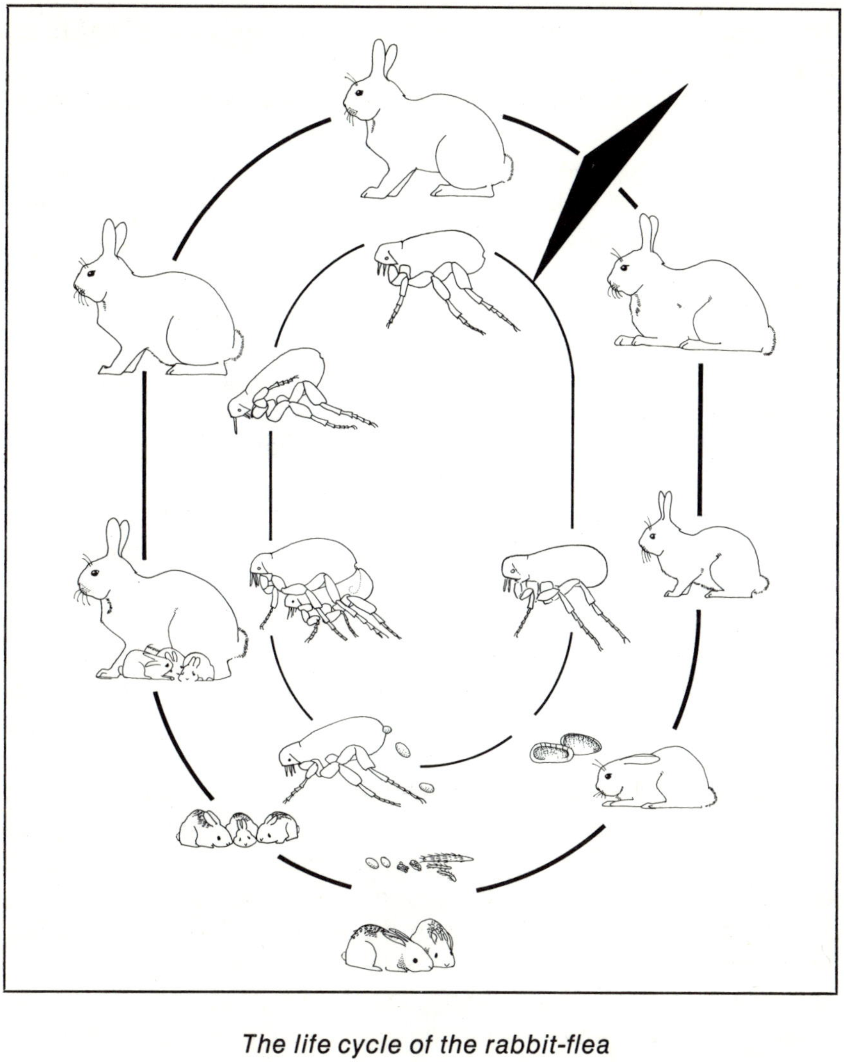

The life cycle of the rabbit-flea

for nearby they will find a constant supply of food in the shape of fine young rabbits. But there is a disadvantage, too: when a virulent rabbit disease like myxomatosis spreads rapidly among the rabbit populations (and myxomatosis is actually transmitted by rabbit-fleas), then millions of rabbits die. Millions of fleas die, too, trapped by their own specialization — for they cannot switch over to different hosts and continue to breed.

There are two other well-known groups of fleas which attach themselves to their hosts. These rather repulsive specialists are stick-fast fleas and jigger (or chigoe) fleas. In both groups it is the females who cling to their hosts while breeding; the males simply feed on them.

Stick-fast fleas are parasites of poultry in tropical and subtropical areas of the world and get their name because of the female's habit of attaching her well-developed mouthparts to a part of the host's skin — usually where it is hard for the host to reach it with its feet — and remaining there until she has laid all her eggs. Infestation with stick-fast fleas can be a serious menace, reducing growth and egg production in poultry and sometimes actually killing off young chickens.

Jiggers are found in parts of South America, Africa, and Asia. Their victims are pigs, dogs, and often humans. Jiggers have favorite spots where they like to penetrate: the feet (often between the toes), the elbows, and the hands. Female jiggers are tiny but their mouthparts are very strong. With the aid of the proboscis they bury themselves under the skin, leaving only the tips of their abdomens exposed so they can breathe through the spiracles and mate. Now — with the female buried — mating takes place and then the male drops off the host. The female is left in her prison; slowly her abdomen swells and so does the victim's skin, perhaps to the size of a pea. The eggs are eventually laid, sometimes several thousand of them, and drop off onto the ground. Then the female dies.

If a jigger is not removed from the victim's skin, serious infections can result and even the loss of fingers or toes. Besides causing sometimes fatal blood poisoning, the female jigger causes intense irritation — a fact which has made many a sufferer "jig" about.

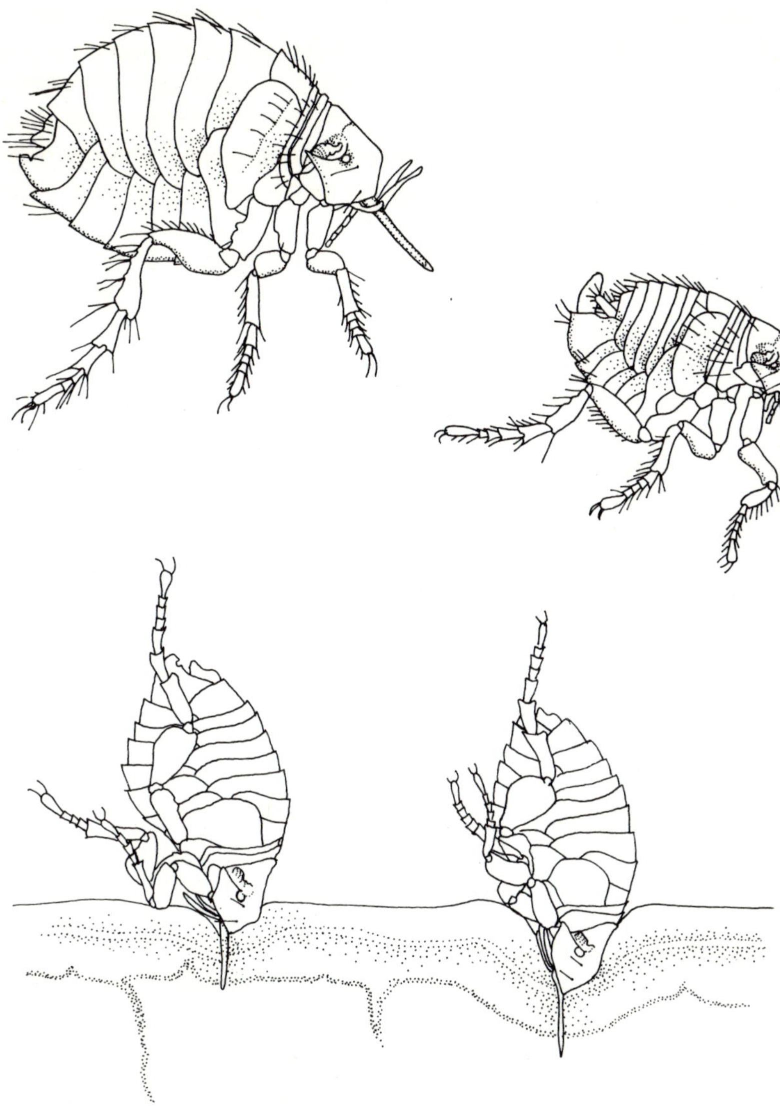

The stick-fast flea: top left, *the female*;
top right, *the male*; bottom, *in action*

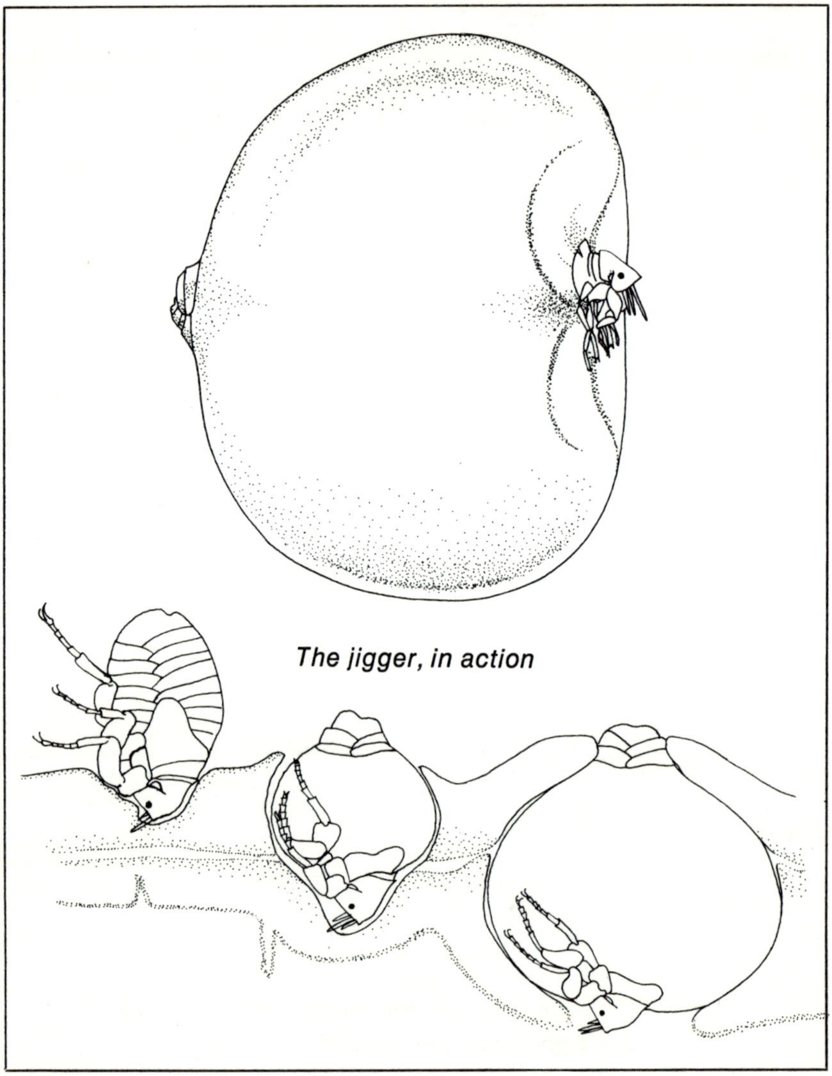

The jigger, in action

15

MATING AND EGG-LAYING

Fleas may have as much trouble finding a mate as they do finding a host. It is quite possible that some die off without producing any offspring simply because they have never had the opportunity to meet a member of the opposite sex. But many, of course, do meet and mate. The millions of fleas in the world are testimony to that.

It is the sense organs, again, which lead a male flea to a female flea. He follows vibrations, air currents, and probably the female's scent. Depending on the species, both male and female fleas may need food — that is to say, blood — before mating. The male, who is smaller and lighter than the female, positions himself beneath his mate, his back to her underside: then clasping her abdomen with his antennae he twists up his tail so mating can take place. The male flea's sexual organs are among the most elaborate in the animal world. His sperm is passed to the female by *two* tiny penis rods, one of which guides the other into the internal organ of the female where the sperm is stored.

For the human-flea mating may take hours but, however long it takes, both male and female may continue to feed at

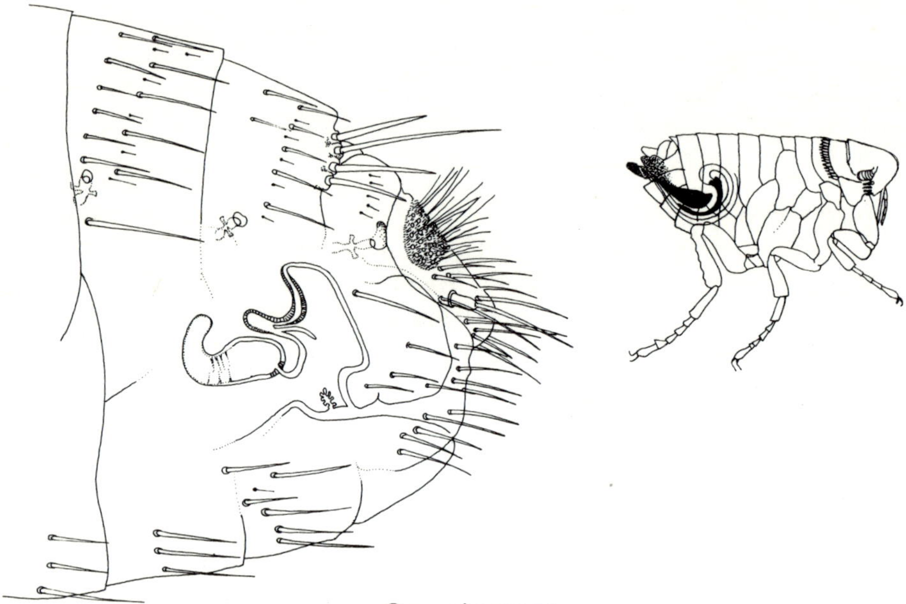

Sexual organs of mouse-flea: upper, *the female*; lower, *the male*

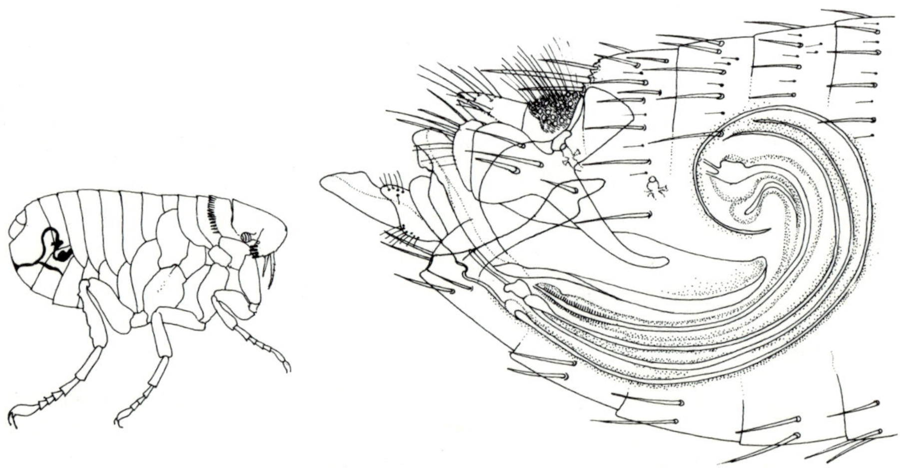

Mating fleas

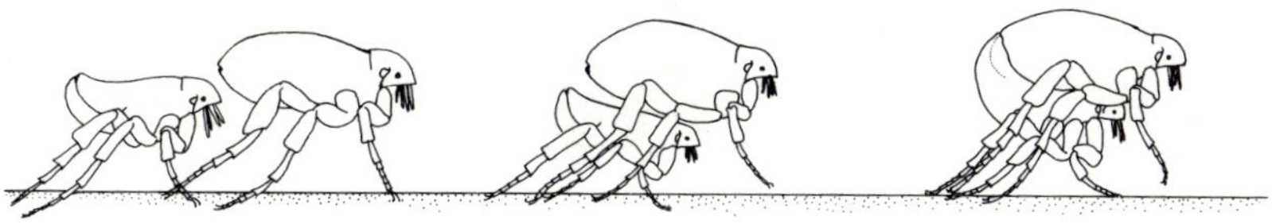

Plague-flea

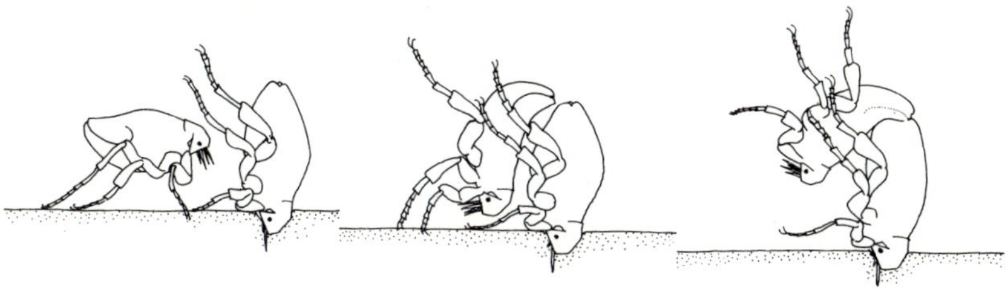

Stick-fast flea

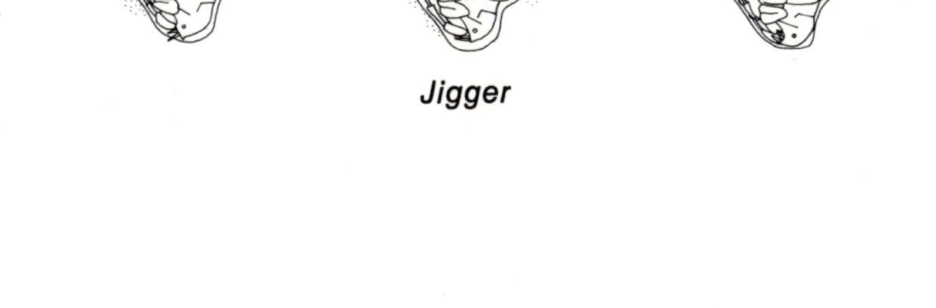

Jigger

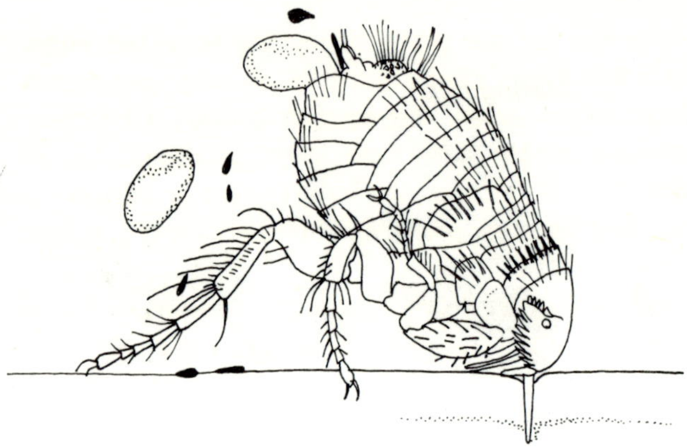

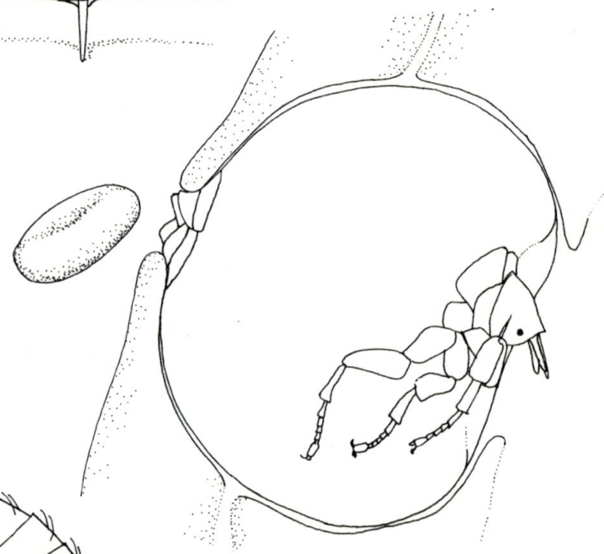

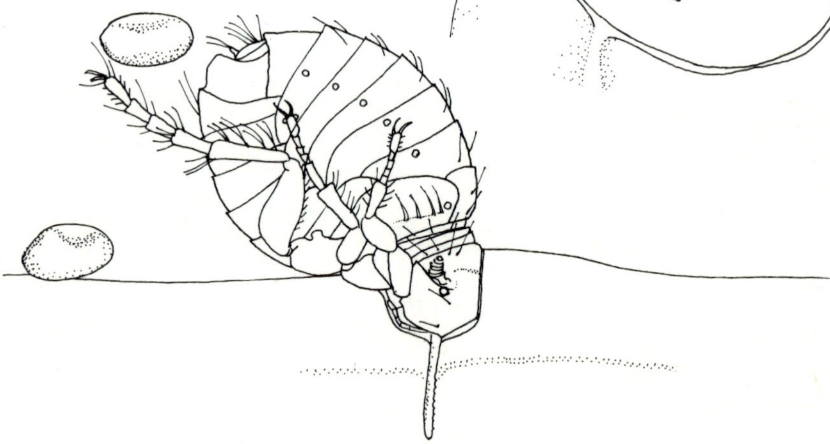

intervals. After a while the female starts to lay the fertilized eggs. Egg-laying may start a few hours after mating, or as long as a month or so later, depending on the species. The eggs drop away in batches. If the female is attached to a host at the time, the eggs usually fall off the host's skin as soon as it scratches and into its nest or burrow. In the case of the human-flea the eggs often fall into cracks and crevices of a house.

The number of eggs laid varies from species to species, but it is usually in the hundreds. One cat-flea laid 284 eggs over a period of four nights. Each morning they were found on the cloth its host had slept on, and counted.

Before each clutch of eggs drops from the female, she feeds, and drops of blood spurt from her anus. In this way the nest area becomes scattered with drops of blood. Strange as it seems this haphazard scattering of blood is of vital importance — it is food for the offspring.

Egg-laying:
top and bottom,
the stick-fast flea;
center, *the jigger*

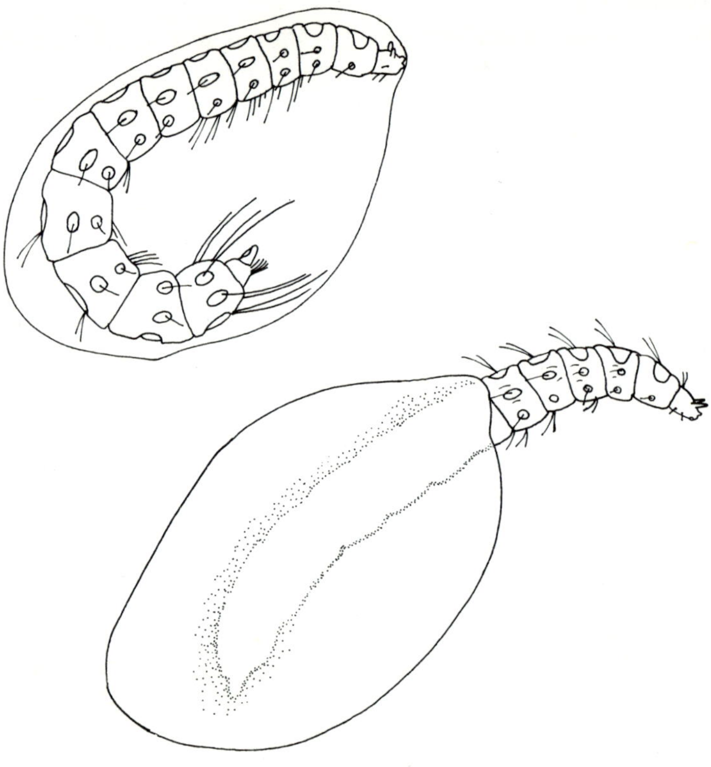

Hatching larvae

16

METAMORPHOSES

There are three "acts" in the development of a flea before it can face life as an adult. Just as being a caterpillar is one stage in the life of the beautiful butterfly, so the immature flea appears in different guises. The changes are known as *metamorphoses.*

The first act in the drama concerns the egg. Fleas' eggs are smooth, white, oval, and easily dented. If they are laid on a dark surface they are visible to the naked eye. After several days or weeks, depending on the species, the embryo (or newly forming flea) is ready to hatch. With a minute "tooth" or "bump" on its head it pushes against the eggshell until it bursts and emerges as a white, wriggling legless creature. The egg has become the *larva*, or grub.

Larvae are divided into fourteen segments, dotted here and there with spiny bristles to help them move about. Although their mouthparts are quite strong, larvae do not *suck* blood and at this stage of their development are not parasites at all. They chew on the dust and dirt and debris which they find in the nesting area and on the scattered droppings of parental blood. Some species cannot develop at all if they do not ingest some blood at

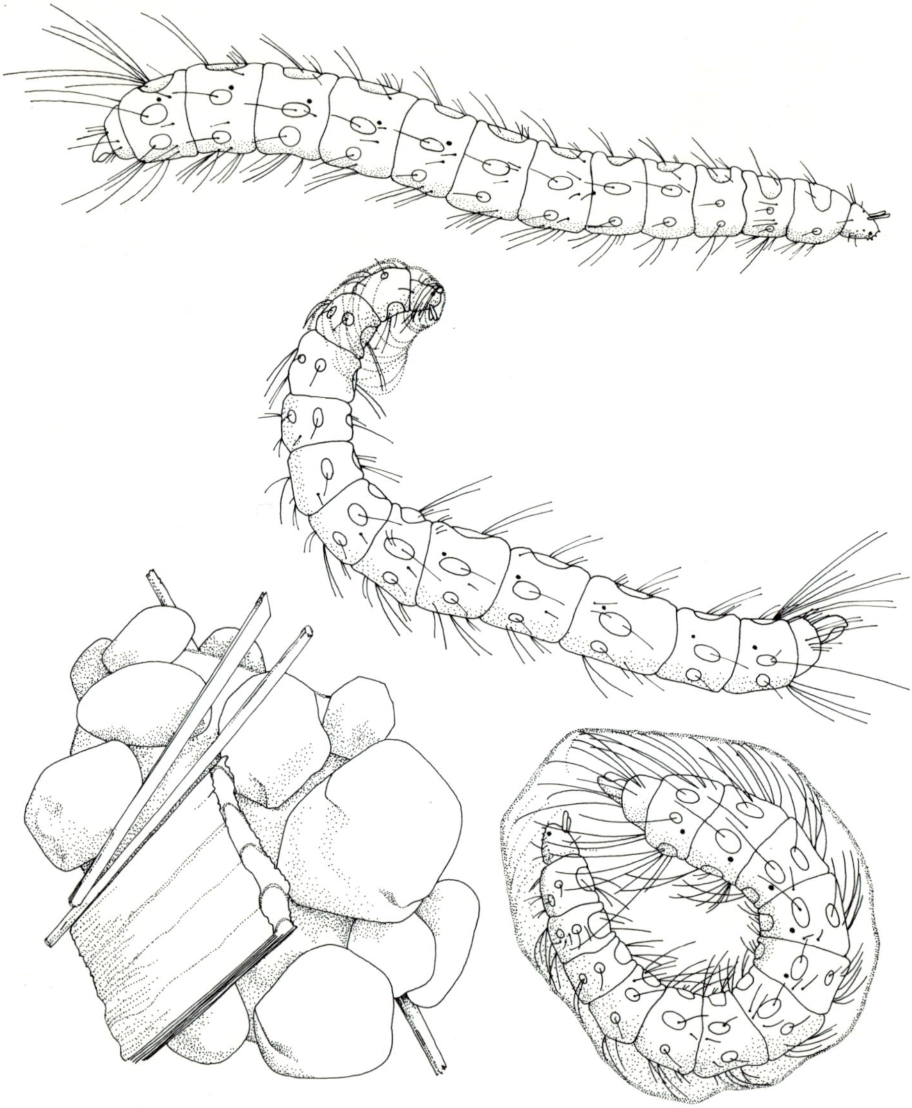

Larva (above); below, *preparation, spinning and camouflage of cocoon*

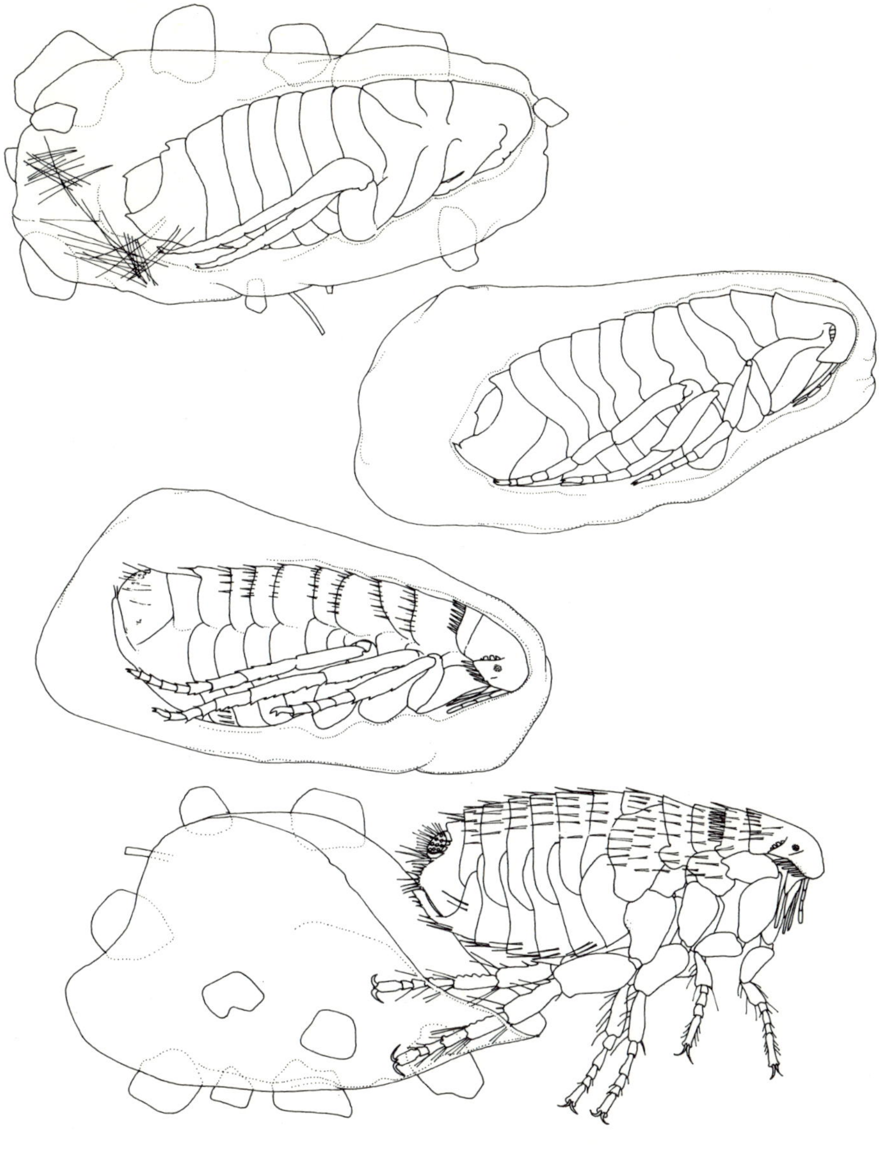

Cocooned and emerging pupa

the larval stage, as they need the iron content for growth. Larvae of human-fleas *can* exist and grow on dust and other organic materials alone. This adaptation helps them to survive in abodes of human beings, which are generally cleaner than the nests and burrows of other flea hosts.

Larvae require food, warmth, and moisture and if they do not get it, they die. If conditions are right the larva will molt twice and after a couple of weeks spin itself a cocoon of silk. The silk is produced by the salivary glands. Now, doubled up in a U-shape, its silk covered and camouflaged by dirt and dust, the larva lies for a couple of days. The scene is set for the third act.

The third stage is the *pupal* stage. The larva sheds its skin and is now wrapped in a thin whitish pupal case, which gradually grows darker. Now it lies dormant. "Pupa" comes from the Latin word for *doll* and many pupae look just like dolls wrapped in swaddling clothes. But beneath their mummified exterior a great deal of work is going on: the adult form is being carefully constructed.

The length of this final pupal stage can vary considerably. It may last a week. It may last as long as a year. The first priority of most species after they emerge from the pupal case and cocoon as adults will be to find a host and feed; there is little point in them emerging if there is no host and no food about. They wait in the cocoon for a clue that there is a host about, for some sort of vibration or stimulus to "awaken" them. The stimulus may come in the form of a bird, poking about in an old nest; in the shape of a rat, returning from a hunting trip; in the shape of a dog or cat back from a "holiday" with its owners. A sudden movement of an animal in the grass, the flutter of a bat as it nears its roost, a spasm of earth turning in a burrow . . . all these can cause a flea to throw off its pupal case and cocoon and start its adult life.

The human-flea may have more trouble "coming to life," for in its case stimulus can be a more unpredictable affair.

17

THE HUMAN-FLEA

Every so often, in the newspapers, there are reports of people being plagued by fleas. Very often, the people concerned have recently moved into a house which has been uninhabited for some time. The new tenants, although they may not realize it, are the "stimulus" that human-fleas need to emerge from their pupal cases. The vibrations of their footsteps and movements and their scent trigger off the final stage of metamorphosis.

Once in Bombay, large numbers of human-fleas were needed for experiments. Some guinea pigs were let loose in a deserted house in the hope that they would act as magnets, for sometimes human-fleas will attach themselves to guinea pigs, pigs, and badgers as well as people. But on this occasion the fleas ignored the bait and none were collected. Then a bare-legged man was sent into the house and in response to his particular scent and vibrations the fleas descended. In one house 150 human-fleas settled on a man within a matter of minutes.

If a vibration comes, human-fleas will burst from their cases in a flash and start to hunt for a host. But if their domain remains deserted they may have to wait for some time (and some have been known to remain alive but inactive in their pupal cases for as long as a year) before commencing adult life.

Once it has become an adult, the flea's life-span is not long. Most species live only two or three months, if that. The most a human-flea can hope for is about eighteen months. But few live that long and only if they have the right combinations of temperature and food supply.

Pulex irritans is the human-flea's scientific name. It means literally "irritating" flea. *Pulex*, or *flea*, may derive from the Latin word *puella* meaning *girl*, because of the flea's supposed preference for female skin. Whether human-fleas do afflict females more than males no one really knows, but it is true that some *human beings* are more afflicted than others. In some way their particular scent is more attractive to fleas.

People often talk about "flea-bites," but the flea does not actually *bite* the skin; it pierces or punctures it with its proboscis. When a human's skin is pierced by the flea's proboscis, it is the flea's saliva and not the wound itself which causes the terrible itching. The actual piercing is rarely felt, though it may leave a dark spot on the skin for a couple of days and the area around it may become reddish and swollen.

Pigs, foxes, and badgers — as well as people — are all hosts to the human-flea. The common denominator is the fact that these creatures live in fairly large, more or less permanent, dwellings. Man probably did not become a host until he changed his earlier nomadic style of living to that of a permanent house dweller.

There are four main places where man is most likely to encounter fleas:

In a house which has been left unused for some time, or in any house where dust and dirt have been allowed to accumulate and so provide ideal conditions for human-fleas to breed.

In the vicinity of pigsties, henhouses, outhouses, and birds' nests.

Around family pets like cats and dogs who may harbor fleas if they are not regularly washed and groomed, or who may "catch" them at the same time they catch other flea hosts — rabbits or mice or hedgehogs.

And around beaches (where the larvae feed on refuse in the sand) and picnic grounds frequented by a lot of people.

In the same way that a flea stands more chance of finding a host in its dwelling place, so people stand more chance of being pestered by fleas if they place themselves in areas where fleas are likely to congregate.

If you are starting to feel itchy, don't become too alarmed. It is quite possible that you will go through your whole life without ever seeing a flea. Since the invention of the vacuum cleaner, which can suck up dirt from all but the smallest crevices, modern sanitation methods, and new ointments and chemical sprays which can quickly cure or wipe out infestations of fleas, they have become much less of a menace to man.

Our forebears were the ones who really suffered.

18

THE SMALLEST SHOW ON EARTH

A pest and a pesterer, yes: down through the centuries. But even so, the human-flea has also given delight and amusement to thousands of people in its role as "artiste"; for along with chimpanzees, dogs, horses, and elephants, the microscopic flea has also known the heady pleasures of stage costumes and circus spotlights. Even now it is the "star" performer in what has been billed as "the smallest show on earth."

Flea circuses may have started as early as the sixteenth century. Right up to the early 1900s they were familiar sideshows, helping to attract crowds to the outdoor markets and fairs. But today, flea circuses are extremely rare — simply because human-fleas themselves are rare in our clean modern homes. There is only one flea circus left in Great Britain and perhaps a few scattered over the continent of Europe and the United States. One of New York's most famous and long-running sideshows, the Broadway Flea Circus, passed into posterity some years ago when its owner, Mr. William Heckler, died.

Mr. Heckler had first made the acquaintance of fleas when he ran away to sea as a young man and served on tramp

steamers. The fleas he caught fascinated him and he began to collect them in bottles. Soon he had worked out a way to "harness" them with tiny gold collars, and chains the width of a hair. Then he began to "teach" them tricks. Mr. Heckler was very particular about which fleas he chose for training. Those who kept trying to jump out of their prisons he dismissed as "stupid"; only the more docile fleas were kept for training. Mr. Heckler's star performer was a flea he found in Ireland and christened Paddy. Paddy was said by Mr. Heckler to have given 52,850 consecutive performances, a record scientists say is quite impossible but which Mr. Heckler liked to call "truly Broadway's longest run."

Along with most other flea circus owners, William Heckler fed all his protégés himself. Human-fleas are used in flea circuses not only because of their spectacular jumping powers, but also because they are easily fed — usually just by extending an arm!

The Tivoli Gardens in Copenhagen was the setting for one of the most famous flea circuses. The star of this Danish troupe was a human-flea which had been caught by a dowager countess on a number 14 tram in one of Copenhagen's most exclusive districts. Resplendent in a pink costume, made from the down of a pedigree flamingo in the zoo, it became the most renowned "rope dancer" of them all. This particular flea had been sent to the circus in answer to a radio appeal; many specimens were needed to keep the circus going. Others were sent from as far away as the United States, but these potential performers were doomed to die in their cold airmail envelopes. Unable to survive at the high altitudes reached by jet planes carrying the mail to Europe, they perished.

19

$2.50 A DOZEN

"Professor" Len Tomlin owns Britain's one remaining flea circus, staged at Manchester's Belle Vue Amusement Park. Life is not easy for the professor — fleas are becoming more and more difficult to come by and sometimes he cannot open his circus at all because his stocks are so depleted. A few years ago he was paying sixty cents for a dozen fleas, now he has to pay $2.50 a dozen. He puts their scarcity down to the fact that so many slums have been demolished in recent years and because people are more hygienic. "People don't sleep on straw mattresses like they used to," he says.

Even at $2.50 a dozen, finding fleas has its pitfalls. The professor and his wife Evelyn, who actually demonstrates the circus, have knocked on doors in the past canvassing for fleas, but people are inclined to take this method personally. Now the Tomlins rely much more on "confidential" advertising.

Professor Tomlin used to be a jeweler and the tools of the jewelry trade — eyeglasses and tweezers — help him now to harness his fleas. Despite his keen eye it took him two years to perfect the art.

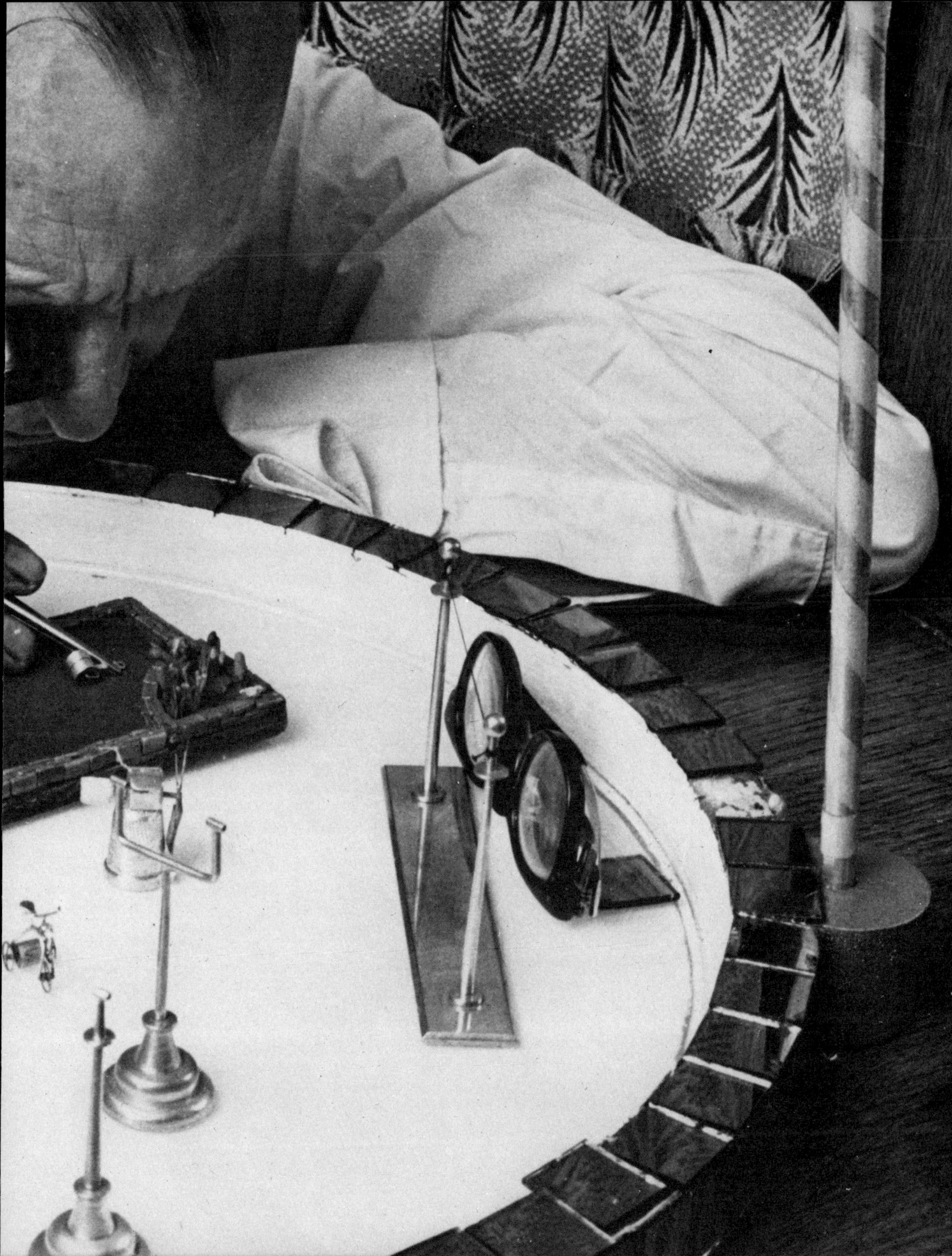

"First of all," he says, "I harness the fleas with a hair-fine silver wire around their 'waists' and 'necks.' Then we have to find out the capabilities of each artiste; some will walk, some will jump, some will hold things. When we have found their capabilities we put them to work, but we must have two of a kind for emergencies."

The fleas in Professor Tomlin's circus "perform" all sorts of acts — tightrope walking, dueling, cycling, juggling, pulling coaches in a "chariot race."

A flea, sorry to say, cannot really be *trained* in the normal sense of the word. It can only react to pulls and tugs on its harness. As the harness is jerked, the flea moves and gives the impression of actually accomplishing such feats as juggling and dueling. And the power packed in their hind legs gives fleas the strength necessary to pull "chariots." Professor Tomlin uses harnesses alone, but in the past it was not unknown for flea circus owners to cement their charges to "carriages," "swords," "orchestra pits," and even to each other, to give a more realistic appearance of dueling, dancing, and musical prowess.

Professor Tomlin's fleas are fed every four hours on his arm, yet even this fondling does not prolong their lives. Most performing fleas live only a couple of weeks. One did achieve a lifespan of a whole month, but this, says the professor, was exceptional. Asked whether he thought it cruel to "train" fleas, the professor's reply was an emphatic *no*. "My greatest ambition is to keep a flea alive, yours would be to kill it — if you could catch one."

20

"MARK WHERE YOUR RIGHT FOOT DOTH STAND . . ."

Flea circuses have been a passing delight to thousands it is true, but fleas have irritated many more people than they have amused. Down through the centuries folktales, proverbs, old wives' tales, rhymes, anecdotes, poems, plays, journals, and travelogs have all harangued them as major pests. Remedies for getting rid of them have varied from tying a patient with a flea in his ear down to a plank, so that his ear hung over the edge, and then striking his feet with a mallet so the shocked flea would jump out; to strewing cow dung over the floors. Susceptible people of the seventeenth century were being quoted this piece of advice: "If you mark where your right foot doth stand at the first time that you do hear the cuckow, and then grave or take up the earth under the same; wheresoever the same is sprinkled about, there will no fleas breed. I know it hath proved true." One can only assume that it did not prove true for many others.

Yet any remedy, patently absurd or otherwise, would have been of little use. Right up until the beginnings of the last century (and in some places for very much longer) the root of the problem lay in the appallingly unhygienic conditions

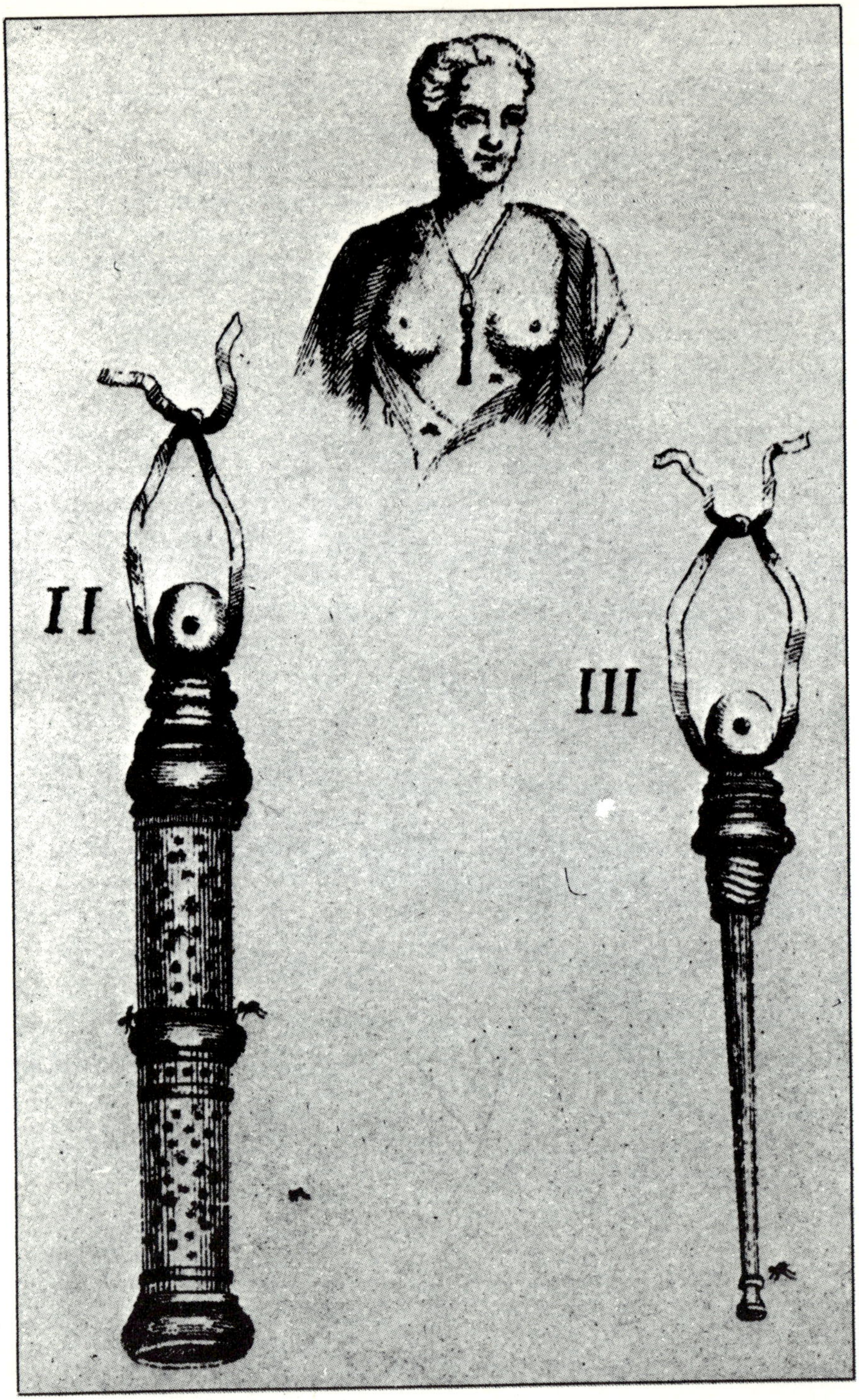
II
III

of houses — and people. It was not unknown for sheep, horses, cows, pigs, and verminous cats and dogs to share living quarters with their owners. Fresh air was considered, by many, to be dangerous and windows and doors were kept shuttered. Baths were thought to be unhealthy and unnecessary, so many of the exquisite clothes of the wealthier people covered filthy louse-ridden bodies. It is really not too mysterious that fleas bred in enormous numbers.

But, then, we have the benefit of hindsight.

A decorative flea-trap women used to wear. Fleas climbed in the outer perforations and were caught on the sticky tube inside

21

SPONTANEOUS GENERATION

We know that fleas are believed to have existed on earth for well over 50 million years. Just how they evolved no one is really sure. But to people who lived before the early seventeenth century the whole field of natural history was hazed in cloud. Evolution and natural selection were words not yet invented and had they have been few would have believed such ungodlike theories anyway.

To the common man, fleas and other small creatures — when he thought about them at all — were just there, pests sent by God perhaps to irk a sinful population. As to their origins, he believed they arose from dirt or dust, sweat or urine, coming to life as the warmth of the sun helped decompose mounds of filth.

This theory of life starting from heat and the decomposition of vegetation was known as *spontaneous generation.*

Some people, ignoring all the evidence that showed reproduction to be the start of new life, believed in this theory right up to the twentieth century.

It was a Dutch biologist, Anton van Leeuwenhoek, who first disproved it. Born in Delft, in 1632, he became fascinated with the microscope, a newly

Leeuwenhoek

invented instrument that was regarded by many scientists of the time as more of a "toy" than anything else.

Leeuwenhoek started to make his own microscopes with small short-focus lenses. Some of his lenses were no bigger than a pin's head, but they magnified an object up to 160 times its normal size, much more than any other microscope yet invented. But to Leeuwenhoek, his microscopes were only a means to an end. What really intrigued him were the minutiae of life — fleas and other insects, blood cells, sperm cells, bacteria. His discoveries included findings on all these things.

Leeuwenhoek did not believe in the theory of spontaneous generation of fleas and he set about proving it. He spent days watching fleas under his microscope and his patience was rewarded, for he was the first to witness all the stages of metamorphoses. He watched as the larvae emerged from the eggs, as they spun cocoons, as they finally burst from their thin pupal cases. He sent letters on his findings — that fleas arose by the normal means of reproduction — to London's Royal Society.

The letters were well received and convinced many. But not all. Still there were those who persisted in clinging to outworn beliefs: but really what did it matter? It was assumed that all there was to know about fleas had now been discovered. There was little more to interest man.

Or was there?

It was not until the last decade of the nineteenth century that fleas were once again brought under the microscope for extended research, and this time they were examined with awe and horror. This time, there *was* more to fleas.

22

THE GERM THEORY

By the early 1890s scientific opinion on disease and its causes had changed dramatically. Louis Pasteur and Robert Koch had proved that a disease which was not caused by a malfunction of the body, was the result of harmful microbes (or germs) and not the act of a vengeful God as many believed. In 1882 Koch had succeeded in isolating the bacteria which caused tuberculosis. Soon, those responsible for diphtheria, cholera, typhoid, and some other epidemic diseases were also isolated and described. Meanwhile, Pasteur was pioneering the use of vaccines, inoculating them into the body to provide immunity to certain diseases.

But there was still some doubt over one major question: how were certain diseases transmitted from one person to another? Answers began to emerge. In 1893 Theobald Smith, a veterinary surgeon, and F. L. Kilbourne, an entomologist, discovered that ticks were responsible for the spread of Texas fever among cattle. A year later Sir Patrick Manson showed that malaria was transmitted to man by mosquitoes.

Patrick Manson

Then, a new threat chilled the world. On January 16, 1894, a case of plague was diagnosed in the city of Canton, China's largest port.

For the third time in history, a major pandemic of the world's most dreaded disease was underway.

23

THE BLACK DEATH

The first record of plague comes in the Old Testament of the Bible (Samuel I, Chapter V). The Philistines, who had defeated the Israelites and stolen the Ark of the Covenant, were afflicted with strange swellings, and "there was a deadly destruction throughout all the city."

The first known pandemic (a pandemic is widespread, covering the world or large areas of it; an epidemic is restricted to a smaller community) occurred during the reign of the emperor Justinian in the sixth century A.D. By 542 A.D. plague was beginning to spread throughout the known world and at the height of the onslaught 10,000 people were dying of it each day — so many, that it became impossible to bury them all.

During the next few centuries a few scattered epidemics were recorded but the pandemic finally burnt itself out. Then, in 1348, the second great pandemic reached Europe. It had begun in Asia and had been carried by traders and sailors to China and India, from where it spread to the Continent and, by the end of the year, it had reached the British Isles. This was the infamous Black Death.*

To much of the terrified and panic-stricken population the Black Death was seen as a sign of God's wrath, a fearful punishment wrought on a wicked populace. No one had any idea of what caused it, though many strange ideas were advanced. Some said it was the result of poisonous fumes escaping from underground; others thought it might be some vile property in the air itself; still others believed it was the work of evil spirits who pricked their victims with poisoned lances or carried the infection from the eyes of the sick to the eyes of the healthy.

For those who contracted it, there was little time to dwell on the meanings or causes; the Black Death was nearly always fatal and many, shunned by friends and relations alike, were left to face their fate alone. In Italy anyone found consorting with a victim of the disease was liable to be burnt at the stake.

A few managed to save themselves. Pope Clement VI spent his time alone in his chambers in France, warming himself between two fires, and lived to tell the tale. An Italian nobleman shut himself and his household, together with plenty of food and drink, behind closed doors and also survived. Few were as lucky. By 1350, when the most catastrophic outbreaks had passed, millions of people had died horrible deaths. One-third of the population of Europe had been wiped out.

This second pandemic continued to flare every now and then right up until the eighteenth century; for a few years it would lie dormant, then rise up again. During one outbreak in 1563, Queen

* The name the Black Death was not used by the people of the Middle Ages. It came into use after the seventeenth century to distinguish it from London's Great Plague of 1665. There was a legend of a black horseman, or black giant, who appeared as a portent of plague and this may be the reason why the name was chosen. Many victims of plague also had black or purplish splotches on their skin.

Elizabeth I retreated with her court to Windsor and ordered gallows to be set up in the market place. Any trader or visitor daring to enter the town from London, where many were dying, was to be hanged immediately without trial or hearing.

London's famous plague of 1665, known as the Great Plague, was more or less England's last epidemic — the parting shot of the second pandemic. The rest of Europe was not so fortunate; outbreaks were still occurring there until the middle of the eighteenth century.

Doctors visiting plague victims believed that their noses should be covered to prevent the spread of infection. This is the costume that doctors wore as late as the eighteenth century when exposed to the plague.

24

THE GREAT PLAGUE OF LONDON

The Great Plague has been called the plague of the poor, for by the summer of 1665 when the disease was at its height, almost everyone else had deserted London. With some exceptions, the royal court, the clergy, the doctors and lawyers, and many of the educated and wealthier classes had hurriedly packed their bags and by wagon and coach and river barge had fled for the country or abroad.

The poor were left unprotected in the stinking overcrowded slums, the reeking alleys, and garbage-strewn streets. As the plague spread from the outparishes into the city of London, red crosses began to appear on doors with the words "Lord Have Mercy On Us" written below. This was the dreaded plague sign.

As soon as a case of plague was reported in a house, it was padlocked and bolted up for forty days and the red cross marked on the door. All the inhabitants were locked in the house, the healthy along with sick and dying relatives. The doors opened only to let out the dead, never the living. If a second case of plague was reported under the same roof, then the period was extended for a further forty days.

A watchman was posted outside each boarded house to make sure no one got out. The watchman kept the keys of the house and sometimes could be persuaded to purchase food and medicines for the families inside. At night, as the clang of a bell signaled the arrival of the deathcart and the carriers bawled their ghastly cry "Bring out your dead," the watchmen unlocked the doors so those who had died could be piled on the carts and carried off to the burial grounds.

At the height of the Great Plague, so many people were dying that the carts were coming through all day long as well as at night, and always — everywhere — were the sounds of mournful clanging bells. As the deaths rose that hot, dry summer, the churchyards became overcrowded and the dead were piled high in pits, their bodies covered with careless shovels of earth.

For the people of that terrible time, life was no more than a candle — one moment flickering, the next snuffed out. In shock and fear they turned to anything that might ward off the disease. Tar and pungent-smelling substances were thought to kill poisons in the air, so they were heaped onto fires in the already airless and shuttered houses. Pipe smoking was thought to be a more personal kind of disinfectant and even school children were encouraged to smoke — one Eton boy was actually whipped for *not* smoking during a morning when it had been prescribed. Doctors and others who were in constant contact with the sick sometimes sucked on lozenges made of dried toads, and one doctor was said to have "cured" himself of plague by keeping a linen cloth, containing a large dried toad, close to his stomach. The toad swelled to an obscene size but no swelling (a sign of the plague) appeared on the doctor.

Mail was thought to be "contaminated" and was sometimes washed and dried in front of a fire before it was opened. In the country letters were often delivered at the end of a long pole. "Contaminated" coins were dipped in bowls of vinegar, common sights outside houses and beside market stalls, before they were exchanged.

Quacks and astrologers were paid handsomely to prescribe remedies and "cures," and all sorts of talismans and charms were carried about. Samuel Pepys, the diarist, carried a hare's foot; some carried

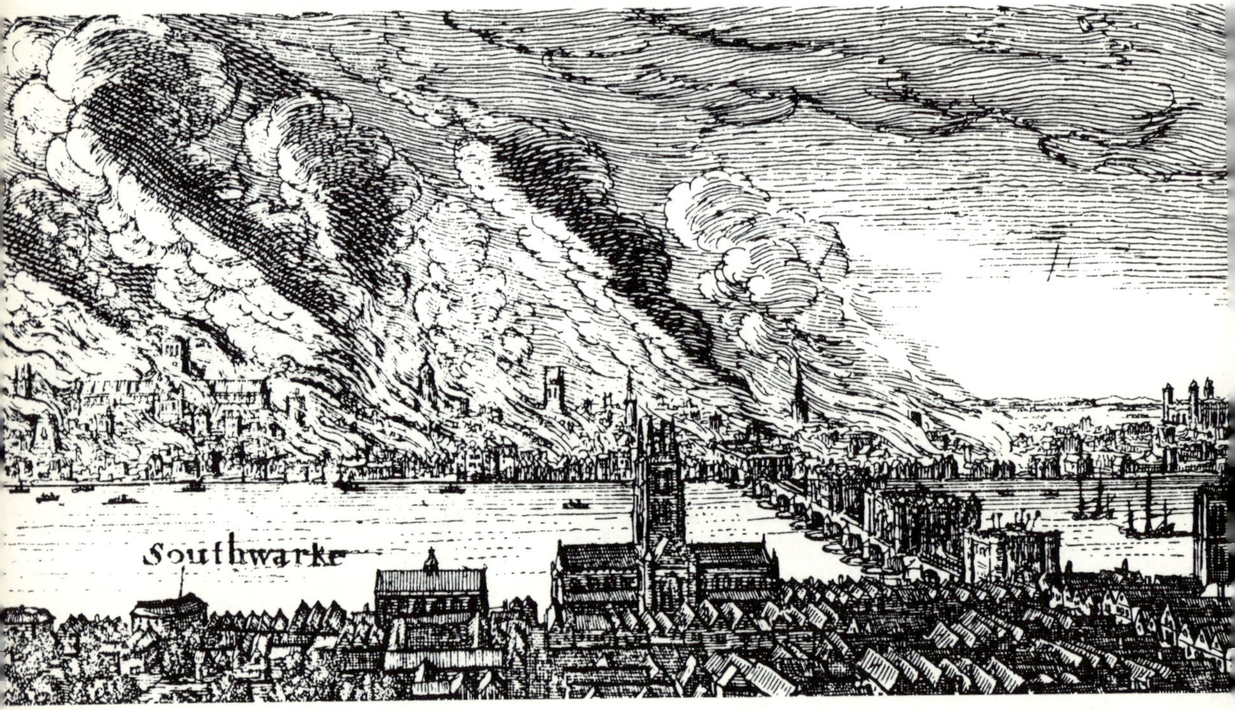
Southwarke

dried toads; others wore sapphires or amber in the belief that sunlight flashing through the stones would draw out infections from their bodies. Certain words were believed to be magical in some way, helping to ward off evil spirits. The word "abracadabra," written in a triangle like this, was often carried about on pieces of paper.

ABRACADABRA
ABRACADABR
ABRACADAB
ABRACADA
ABRACAD
ABRACA
ABRAC
ABRA
ABR
AB
A

Yet, nothing worked. The deaths mounted and by the end of the year something like 100,000 people had died agonizing deaths. When the Great Fire of London swept through the city in the following year (1666) and destroyed five-sixths of it, it also helped to wipe out the conditions which encouraged plague . . . the cramped, timbered houses; the slums; the filthy narrow alleys; the great mounds of garbage; the totally inadequate sewerage system.

All favorite haunts of rats and rat-fleas.

Whenever plague broke out, people abandoned the disease-ridden cities and fled to the country.

The Great Fire of London, 1666, as seen from south of the River Thames

25

A SHORT STEP TO MAN

Plague is primarily a disease of rodents. In some parts of the world — Asia, Africa, and South America — plague bacilli (rod-shaped bacteria; *singular,* bacillus) exist in the bloodstreams of marmots, squirrels, and various wild rodents without doing much damage. But if those infected wild rodents come into contact with more domestic species of rats, as they do from time to time during migrations or changes in weather conditions or harvests, then there is a good chance that the rats will succumb to the diseases. In time, the bacilli will be transmitted from rats living in outlying districts to town-dwelling rats. From there, it is but a short step to man.

The one constant factor in all these transactions is the flea. Rat-fleas — and particularly a tropical rat-flea, whose scientific name is *Xenopsylla cheopis* — transmit plague not only from one animal to another but also from rodents to man.

As soon as a rat dies (and thousands and probably millions do during times of plague) its body grows cold. Now its parasite, the rat-flea, has to look around for another host. It is at this point that man can become infected with plague if he is the nearest warm-blooded animal about.

Man catches plague when his skin is pierced by a rat-flea who is suffering from the disease. Although bacilli clinging to the flea's proboscis can be transmitted into the wound, they are normally literally "vomited" into the wound. In fleas of certain species, which have been feeding on plague-infected rats, the plague bacilli multiply to such an extent that their stomachs become blocked with them. The starving fleas try to feed, but fresh blood cannot pass into their stomachs. Soon, they are so congested with infected blood that some of it is forced back down the proboscis — or "vomited" into the wound. Then the bacilli multiply in the new victim's bloodstream.

Man is heir to three types of plague. Bubonic plague is the most well known. The patient becomes feverish and in his delirium may start staggering about. Then a swelling (and it is usually only one) appears in the groin or neck or under the arm. The name bubonic comes from the Greek word *bubon*, meaning *groin.* The swelling, or "bubo," may be the size of a nut or the size of an orange and until it begins to break down and discharge, there may be agonizing pain. Some swellings do not discharge but just disappear after a while. In the past, most cases of bubonic plague were fatal; now, while still extremely serious, many cases can be cured with new drugs.

Septicaemic plague is also carried by fleas. In this type of plague the brunt of the infection lies in the bloodstream and the patient may die in a matter of hours. Doctors during the Great Plague of London were baffled by this form of the disease; there were no swellings, and a person who had been cheerful and healthy one day, could be suddenly dead the next.

Pneumonic plague is a massive infection of the lungs. Each time the patient coughs or sneezes the bacilli are sprayed into the air and spread to someone else, in the same way as influenza. Fleas play no role in the *transmission* of this disease. Both septicaemic and pneumonic plague can be fatal if they are not diagnosed quickly, but now they are curable with antibiotics.

Plague is, of course, fatal to the fleas and the domestic rats who carry it, as well as to man. A starving flea with its stomach blocked will probably die within a week. A rat with the disease will succumb in a few days. But it was not until the end of the last century that anyone realized that the deaths of large numbers of rats could also herald death and disaster for millions of human beings.

26

THE THIRD PANDEMIC

When the daughter-in-law of a Chinese General in Canton fell ill with plague in January 1894, no one could have guessed that this was the start of a third pandemic. The most catastrophic of them all, it would eventually spread to Australia, South and Central America, and parts of the United States, South Africa, and other parts of Africa and the world, and would take the lives of over twenty *million* people.

By the end of March the disease had reached epidemic proportions among the poor inhabitants of Canton. Again, like the victims of London's Great Plague, the poor were the most afflicted, for it was in their overcrowded, badly constructed houses that rats throve. Thirteen dead rats were swept from one house in a single morning.

By May, plague had spread to the crowded city of Hong Kong and, as the horrible symptoms became prevalent, scientists were sent to the city to study the disease. Two men, Professor S. Kitasato and Dr. A. Yersin, working independently, both discovered and isolated the plague bacillus for the first time.

Hong Kong was, and is, a busy port. By 1896, rats stowed away on ocean-going vessels had spread the disease to the Indian port of Bombay. The world trembled at the thought that soon plague might be rife around the globe. Various countries, including Great Britain, Russia, Italy, Austria, and Germany, hastily dispatched commissions to study plague at first hand in Bombay. But there were so many commissions, engendering so much confusion, that it soon became a case of "too many cooks . . ." It was the independent scientists, working on their own, who came up with the most significant data. One such scientist was a Frenchman, P. L. Simond.

Simond observed that most human cases of plague occurred where rats were most numerous — around granaries and warehouses. At one mill, twenty men had cleared away a quantity of dead rats before starting work one morning. Within three days half the men had contracted plague. At a hotel in Calcutta, one of the guests had a pet dog who arrived back one day with a dead rat in its jaws. The dog dropped the rat on the man's bed and later the guest fell ill with plague.

Up to this time only a few people had imagined there was any connection between plague and rats. During London's Great Plague, for instance, thousands of dogs and cats had been slaughtered when it was suggested that they might be responsible, but few people had thought of trying to eliminate rodents. Perhaps only the Orientals had ever really believed in the connection; many of their ancient traditions warned against touching a sick animal.

Slowly the idea that plague was primarily a rodent disease grew. And then Simond began to notice that plague-infected rats seemed to carry more fleas than healthy rats. Surely, thought Simond, these fleas were likely to become infected themselves? To test this theory he fed fleas from a healthy rat on a plague-infected mouse. Twenty-four hours later he observed plague bacilli in the stomachs of the fleas.

Most important of all, Simond noted that, given the opportunity, rat-fleas would feed on man. Was this the way the disease was transmitted? Members of the various commissions reacted strongly against Simond's theories. They held the firm belief that fleas did *not* change hosts, and certainly not from rats to man. But some months before, another scientist, Ogata of Japan, had put forward the same idea and doubts began to crumble. Both Simond and Ogata were proved correct a few years later when a man named W. Glen Liston conducted experiments which confirmed their theories. Liston showed that people who worked in dust-ridden places, in granaries, warehouses, and so on, were more inclined to catch plague than others — for these were the areas where rat-fleas were most likely to breed. Eventually, in 1905, the Plague Research Commission was set up to organize findings and try to sort out all the confusion. Liston was made a senior member.

The commission examined thousands of dead rats and conducted experiments with living rats and rat-fleas. It proved that when both sick and healthy animals were put in specially built compartments from which fleas were excluded, plague did not spread from the sick animals to the healthy ones.

In one experiment guinea pigs were placed in cages suspended some two feet from the ground. In the cabin around them were many fleas and many plague-infected animals, but the guinea pigs did not contract plague. Once their cages were lowered to within inches of the ground, however — that is, within the height that fleas can jump — the guinea pigs became infected.

By 1910 when the commission ended its work, no one was left in any doubt that fleas were the transmitters of plague. Within a few years, more had been learned about man's most feared disease than in the whole of the previous era of human history.

27

PLAGUE IN THE TWENTIETH CENTURY

In the first forty years of the third pandemic, there were fewer than one hundred cases of plague reported in Britain. The rest of Europe suffered very little too. Why did Europe escape the brunt of this pandemic, when it had suffered so much in the past?

Many experts believe that during the eighteenth century when the last major epidemic of plague occurred in Europe, there was a change in rat and rat-flea species. In 1727 hordes of brown rats (*Rattus norvegicus*) crossed Russia's Volga River and spread out over the continent and into England, driving out the much more domesticated black rats (*Rattus rattus*). Black rats live in close proximity to man and are hosts to, among others, the most effective plague-flea *Xenopsylla cheopis.* Brown rats are not as domesticated as black rats and inhabit warehouses, dock areas, and sewers much more than they do the homes of man. They are hosts to different species of fleas, the most predominate of which is a species known as *Nosopsyllus fasciatus* which does not have much affection for human blood. Because it does not change hosts so

Top, *the brown rat*;
bottom, *the black rat —
carrier of the plague-flea*

readily, scientists believe that *N. fasciatus* cannot play much part in spreading plague among humans.

During the third pandemic and the epidemics that have occurred in its wake, plague has been mainly a tropical and subtropical disease, for these are the areas of the world where most black rats are now found.

Still, human beings have taken no chance that plague might break out again. Ratproofing of ships has been enforced, sanitary inspections tightened up, and new insecticides and vaccines developed. During World War II plague could very easily have become widespread. There were extensive starving rat populations and conditions of filth and famine in many areas. The outbreaks, in fact, were minimal. Precautions were taken to keep ships and planes free from rats; D.D.T., a potent insecticide, was used to disinfect premises and people; and immunization programs and enforced quarantine of vessels helped keep the disease at bay.

28

"GREAT FLEAS HAVE LITTLE FLEAS . . ."

Unwittingly, fleas have caused as much misery as a Hitler or a Stalin. Besides spreading plague, they can also transmit other diseases such as myxomatosis to rabbits, and murine typhus and a plague-like disease called tularaemia among rodents and occasionally man. The latter two are quite widespread, but can be cured.

As we noted at the beginning of this book, fleas have a lot to answer for. Yet fleas do not have it all their own way; they, too, have their enemies. Ants and beetles eat fleas. So do birds. And fleas' hosts, with their sharp claws, will scratch and destroy them. Man's fingernails are effective pincers, too.

And just as fleas are parasites of some warm-blooded animals, fleas, too, have parasites. There is a famous rhyme about them:

Great fleas have little fleas upon their backs to bite 'em,
And little fleas have lesser fleas, and so ad infinitum.
And the great fleas themselves, in turn, have greater fleas to go on;
While these in turn have greater still, and greater still, and so on.

The "little fleas" of this rhyme are mites. They may use fleas purely as a means of transport, usually without causing any harm. Tiny wormlike organisms are also parasites of fleas. They feed on the gut and other internal organs, sometimes so voraciously that the fleas die.

What is the point of these parasites?

What, indeed, is the point of parasitic fleas? Their main job is despised but necessary. Parasites are population-checkers, nippers-in-the-bud of over-abundant life. Along with predators, tornadoes, diseases, and fluctuating food supplies, they are natural influences keeping nature's balance of numbers. For without some sort of check on their growth, animals and plants have the potential to swamp the earth.

So, in the end, perhaps there is something to be said for the existence of fleas.

BIBLIOGRAPHY

Askew, R. R. *Parasitic Insects.* New York: American Elsevier Publishing, Inc., 1971.

Bells, Walter G. *The Great Plague in London in 1665.* New York: Dodd, Mead & Company, 1924.

Cushing, Emory C. *History of Entomology in World War II.* Washington, D.C.: Smithsonian Institution, 1957.

Defoe, Daniel. *A Journal of the Plague Year.* New York: E. P. Dutton & Co., Inc., 1953 (Originally published in 1722)

Doane, Rennie W. *Insects & Diseases.* London: Constable & Company Ltd., 1910.

Hirst, Fabian L. *The Conquest of Plague.* Oxford: Clarendon Press, 1953.

Lapage, Geoffrey. *Animals Parasitic in Man.* New York: Dover Publications, Inc., rev. ed., 1963.

Lehane, Brendan. *The Compleat Flea.* London: John Murray (Publishers) Ltd., 1969.

Oldroyd, Harold. *Elements of Entomology.* New York: Universe Books, 1970.

———. *Insects and Their World.* Chicago, Ill.: University of Chicago Press, 1960.

Owens, G. and E. Browning. *Some British Mites of Economic Importance.* Pamphlet. Economic Series No. 17. London: British Museum (Natural History), Department of Zoology, 1955.

Rothchild, Miriam and Teresa Clay. *Fleas, Flukes and Cuckoos: A Study of Bird Parasites.* London: William Collins Sons & Co. Ltd., 1952.

Russell, Harold. *The Flea.* Cambridge: Cambridge University Press, 1913.

Shipley, Sc.D., A. E. *The Minor Horrors of War.* London: Smith, Elder & Co., 1915.

Smit, F. G. A. M. *Fleas.* Pamphlet. Economic Series No. 3A. London: British Museum (Natural History), 1958.

———. *Lice.* Pamphlet. Economic Series No. 2A. London: British Museum (Natural History), 1942.

Ziegler, Philip. *The Black Death.* New York: Harper & Row, Publishers, 1971.

Zinsser, Hans. *Rats, Lice & History.* Boston: Little, Brown & Company, 1935; New York: Bantam Books, Inc., 1960.

INDEX

ABOUT THE AUTHORS

Helen Hoke's first job was with her father's country newspaper in a small Pennsylvania town. Since then she has been a teacher, a book-shop manager, an editor, and a prolific writer of books for children and adults. *Fleas* is her seventy-first book.

She is married to publisher Franklin Watts. They live in London, England, where she continues to write and edit from a house with a lovely roof garden.

Valerie Pitt was born and educated in England and received a diploma in journalism from London Polytechnic. She has been a reporter and writer for several English newspapers and magazines. At present she lives in London where she writes a book review column for an English monthly. She is the author of a number of children's books.